Industrial Automation and Information Technology

Michael Weyrich

Industrial Automation and Information Technology

IT Architectures, Communication and Software for System Design

Michael Weyrich
Stuttgart University
Stuttgart, Germany

ISBN 978-3-662-69245-5   ISBN 978-3-662-69243-1   (eBook)
https://doi.org/10.1007/978-3-662-69243-1

Editorial Contact: Michael Kottusch
This Springer Vieweg imprint is published by the registered company Springer-Verlag GmbH, DE, part of Springer Nature.
The registered company address is: Heidelberger Platz 3, 14197 Berlin, Germany

Preface

Automation and information technology drive innovation and job creation across various industrial sectors. As a result, there has been a significant increase in the demand for lectures on industrial automation technology. The Institute for Automation Technology and Software Systems in Stuttgart, Germany, offers courses on automation technology to cater to this demand.

In the past, numerous discussions with students and professionals have shown that combining methodological and practical knowledge is crucial for understanding the technical aspects and complex relationships inherent in using automation technology in real-world scenarios. This blend of expertise is essential for understanding and applying the technical aspects and concerns of IT and software in automation.

The knowledge presented in this book is intended for a wide range of readers, including undergraduate and graduate students, engineers, technicians, scientists, managers, and decision-makers who are active in, or interested in, automation and the use of industrial information technology. In this age of advanced technology such as artificial intelligence and the Internet, it is crucial to handle information accurately and attentively. Despite the numerous sources available, such as search engines, chatbots, and language models, it can still be challenging to try to obtain a comprehensive understanding of automation technology. This knowledge is vital for asking the right questions and conducting successful research in the Internet.

Although it took longer than expected to create and compile the material, I am happy to be in a position to meet the request for a textbook on industrial automation and information technology. The book provides a concise combination of broad and deep knowledge that illuminates IT for automation without getting bogged down in the enormous depth and complexity of today's technologies. It presents a comprehensive overview of current and future technologies and their applications in automation along with real-world case studies.

The book is now available in English due to the increasing internationalization of the German education system and the many international cooperation schemes of German companies worldwide. The aim of the book is to introduce the most important topics with regard to IT for automation technology while considering the ways of thinking which characterize German engineering and the concepts of Industrie 4.0. It presents research results descriptively for teaching purposes and includes empirical knowledge and observations gleaned from years of industrial work carried out in international electrical engineering and automotive companies, industrial research projects, and academia. Insights gained from my position as chairman of the Society for Measurement and Automation Technology within the Association of German Engineers (VDI/VDE) and on the board of the German Association for Electrical, Electronic & Information Technologies (VDE) have helped me prioritize the many topics.

Due to the wealth of information available, a concise presentation of the subject is necessary in order to provide an introduction to the complex world of automation. The book consists of three sections: fundamentals and essential technologies, case studies of applications, and aspects of value creation as well as practical uses. It provides essential skills for developing concepts and principles to be found in the field of information technology for automation and applying them to the areas in which they are used. It also addresses the economic contexts important for planning, realizing, and implementing information technology for automation.

After reading this book, readers will be able to acquire specific in-depth knowledge of this rapidly developing field with the help of further reading, presentations in the Internet, or training courses on industrial products. On this basis, AI chatbots using natural language processing models can help readers formulate the right questions.

I would like to express my gratitude to the numerous contributors to this book for the research they conducted and the sub-aspects they illuminated.

Benjamin Maschler and Dustin White worked hard on the didactic presentation of the material and provided concrete feedback on individual chapters. We should also mention Matthias Klein and Philipp Marks, who were very active in the initial research concerning the current scope of the material. During the preparation of the case studies, Dominik Braun, Falk Dettinger, Daniel Dittler, Baran Gül, Andreas Löcklin, Nada Sahlab, and Matthias Weiß contributed to the design drafts in particular. Dr. Nasser Jazdi and Falk Dettinger reviewed the manuscript with care and clarified any technical discrepancies that came up.

Ulrike Bek provided extensive assistance in creating the graphics for the blocks; one image is the work of Timo Müller, whereas Justina Trefz and Oliver Schneider were responsible for the Photoshop illustrations. I would like to thank Dr. Susanne Schuster for her final manuscript review and the advice she gave on structure and readability as well as Paul Harrison for reviewing the English version.

I would like to inspire as many students as possible to immerse themselves in the fascinating field of information technology and software in the automation industry. With this in mind, I wish the readers of the book a wealth of stimulating discoveries and insights.

Stuttgart, Germany   Michael Weyrich
April 2024

Contents

Introduction 1

Abstract

This book regards automation from the perspective of IT and software. It outlines the key elements of various solutions and describes how IT-oriented automation systems can be designed today and in the future.

This brief introduction provides an overview of what automation technology entails and which information and skills the book covers. For this purpose, it discusses the following questions:

- What is our understanding of automation technology, and which topics do engineers feel responsible for?
- What is the composition of the automation technology stack described in the book?
- How is the book structured, what are its learning objectives, and what competencies does it teach?

Due to the wide variety of technical solutions, the focus is on understanding the process of creating complex automation and not on providing in-depth knowledge of specific procedures and methods. The reader will learn how complex automation systems can be implemented and which criteria form the basis for design decisions.

1.1   What Automation Technology Accomplishes Today and in the Future

Automation technology has been, and will continue to be, one of the driving forces behind progress, bringing about changes in business processes, industrial workflows, and new products designed to improve our living and working environments.

In a forward-looking study for the year 2025, the Society for Measurement and Automation Technology within the Association of German Engineers (VDI/VDE) outlined a number of theses for the development of automation technology that still hold up today (Adamczyk et al. 2015):

- Automation enables flexible value networks.
- Automation makes it easier for people to use technology.
- Automation is the integration of different technologies and the interconnection of various disciplines.
- Automation connects real-world physical elements and their digital representations.

Automation technology plays a vital role in handling the constantly growing complexity of technical processes and creating intelligent networked products. The focus is on exploring new functional principles arising from technological opportunities and creating concepts, systematic analyses, and experimental setups allowing us to investigate the wide range of issues involved.

For instance, many companies are evaluating their digital maturity in terms of Industrie 4.0 or assessing their capacity for digital transformation. However, intelligent products—systems that develop abilities on the basis of networked data, that is—are increasingly becoming an essential topic for corporate executives, who need to consider how novel product capabilities can be developed on the basis of data and software.

Automation technology will continue to significantly transform our world and the way we work.

As Porter and Heppelmann (2014) put it:

Ultimately, (smart) products can function with complete autonomy … human operators merely monitor performance or watch over the fleet or the system rather than individual units.

The development of automation technology raises many questions about economic exploitation and social acceptance which cannot be left solely to the companies focusing on technical possibilities.

Insights into these technologies and the mechanisms for the successful use of automation technology in applications are highly diverse and are subject to influencing factors and accompanying circumstances that have far-reaching consequences for society.

In Europe, the regulatory framework with its standards and integration approaches for artificial intelligence systems is under intensive discussion. It is seen as a fundamental requirement for the creation of a digital ecosystem. Industry associations and federations of industrial companies are particularly active in Europe, with their technical experts addressing overarching issues.

Future fields of application range from the automation of knowledge work to robotic applications and the further rationalization of industrial value chains and production. Good working conditions and qualified training will be the cornerstones of technology development.

Automation technology also enables sustainability and a circular economy. Virtual data spaces, for example, enable us to collect information about product manufacturing such as the carbon footprints of products, creating a new form of transparency that provides deep insights into the value chain, its actors, and business relationships. However, it must be realized that the storage of data and the computing power of automation technology require a significant amount of energy.

The impact of information networking and the possibilities of artificial intelligence are only beginning to become apparent, and much more discussion and development are needed before generally accepted, sustainable guidelines and rules can emerge, and these may need to be regulated by the government. In all of these discussions, however, technologies are the catalysts for change and the shaping of our world. The consistent use of software and AI and the use of interlinked data and information are vital for automation technology.

1.2 Automation Technology's Self-Perception

Automation technology is a multidisciplinary field that combines mechanical engineering, electrical engineering, computer science, and information technology. It plays a crucial role in system integration by connecting system elements and coordinating their interactions.

Automation experts view themselves as system integrators responsible for organizing technical processes, simplifying complexity, and designing entire systems to work seamlessly.

Over the last five decades, automation technology has had a significant impact on company strategies and international competition. It has increased the productivity of industry and enabled technical products to acquire new functions and capabilities.

Automation technology has evolved over the years, and limited computing power has made real-time control and essential communications vital in the past. Today, automation technology focuses more on the omnipresence of information technology in networked data collection, which enables decision-making by deriving courses of action from data and software analysis.

It is possible to automate any area, and it is becoming increasingly common to combine small- and large-scale automation. Additionally, the dominance of software that can be modified virtually over the air anytime and anywhere has become more prevalent. We now

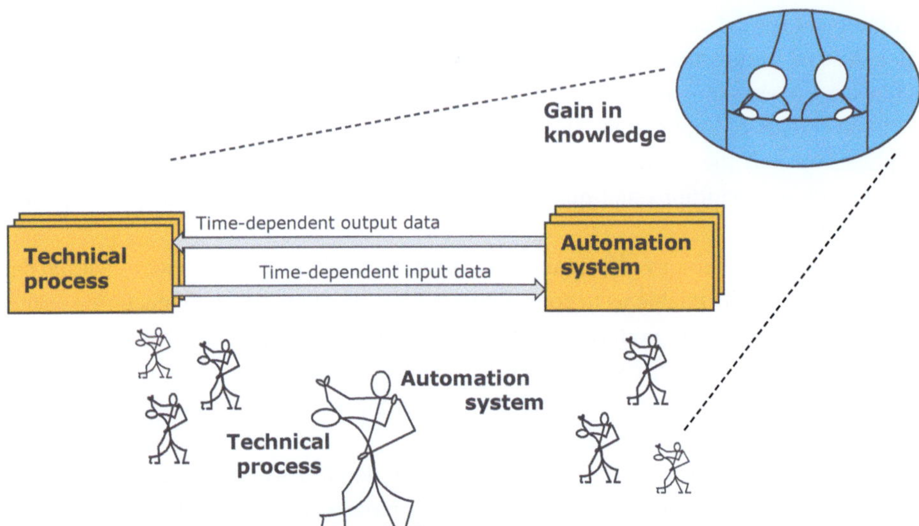

Fig. 1.1 An analogy to illustrate the connection between the technical process, the automation system, and the possibilities of higher-level knowledge acquisition (inspired by an illustration by Rudolf Lauber, 1989)

speak of "software-defined systems," which are not determined by the particular structure of the hardware of the technical process but by the configuration of the automation software.

The automation industry has come a long way, and Fig. 1.1 depicts today's automation technology and its key topics. The dancing couple illustration used by Professors Rudolf Lauber and Peter Göhner in their textbooks (1989, 1999) visually represents how automation systems control and regulate complex processes in unison in a way similar to dancing. At that time, the focus was on real-time capability, meaning the timely and synchronized coordination of technical processes, and this faced many challenges.

Automation technology has grown increasingly complex over time. While technical processes still require coordination, modern automation includes a far broader range of tasks. The image of the dancing couple symbolizes the growing intricacy of automation engineering in modern times. It exemplifies the diverse elements that need to be seamlessly integrated to form a complete end-to-end system. In terms of overall synchronization, it entails overseeing multiple couples dancing in unison rather than directing the movements of one couple alone. This means the orchestration has taken on a new meaning, incorporating experiences from other dance schools and requiring more networking.

Real-time control of automation systems remains essential, but new technologies have expanded its scope. Today, the focus is on coordination with other systems, interconnectivity, and overall assessment. In terms of our dancing couple metaphor, this means that the style of music may change, so that people may be dancing to a different rhythm tomorrow than they are today.

Coordinating a single automated system can still be challenging, especially for new automation tasks. However, orchestration and optimization of several approaches can help create complex, resilient, and agile systems suitable for ever-changing tasks.

The example of the dancing couple illustrates the complexity of automation technology and the need to integrate various aspects to create an end-to-end process.

In many industrial applications, reliability, resilience, energy efficiency, and the cost of change involve more than just implementing a specific function in real time. Automation systems and their components are monitored at a higher level, providing insights into their behavior and overall context.

Automation is now an interdisciplinary field connecting production owners and manufacturers of equipment and systems with vendors of automation and IT. Together, they make automation happen and act as the "glue" that holds the individual players and their contributions together.

The future of automation technology will see autonomous systems set new standards. The challenge lies in how to perceive the environment and to which extent an autonomous system with more freedom of action can influence it. By combining sensory inputs and cognitive abilities, autonomous systems can understand problematic situations and solve them on their own.

The trend in automation technology is moving towards systems with automatic or autonomous capabilities which are interconnected, perceive their environment, and act autonomously.

1.3   Automation Technology in this Book

This book aims to provide readers with a comprehensive understanding of automation systems. After clarifying certain key terms, it presents central automation engineering technologies from an IT and software perspective. This approach enables the reader to become familiar with the fundamental technologies, methods, and procedures.

The book then uses case studies to discuss the contextual circumstances under which the integration of automated systems occurs and the foundation upon which decisions for or against specific technologies are made. It elaborates on various automation case studies and includes sensor technology, industrial production, and aspects of automotive IT.

The book concludes with a discussion of non-technical aspects of value creation. It covers the realization of value creation throughout a product's lifecycle in addition to technical methods and process expertise. It also provides a topology of industrial automation and information technology as shown in Fig. 1.2.

Lastly, the book discusses technical products and processes in the context of the business perspectives and technology deployment of automation technology.

The field of automation technology is vast and intricate, encompassing various technical concepts and tools utilized to develop and manage industrial processes. Its technology stack is extensive and includes a range of technologies at different levels of maturity. Some

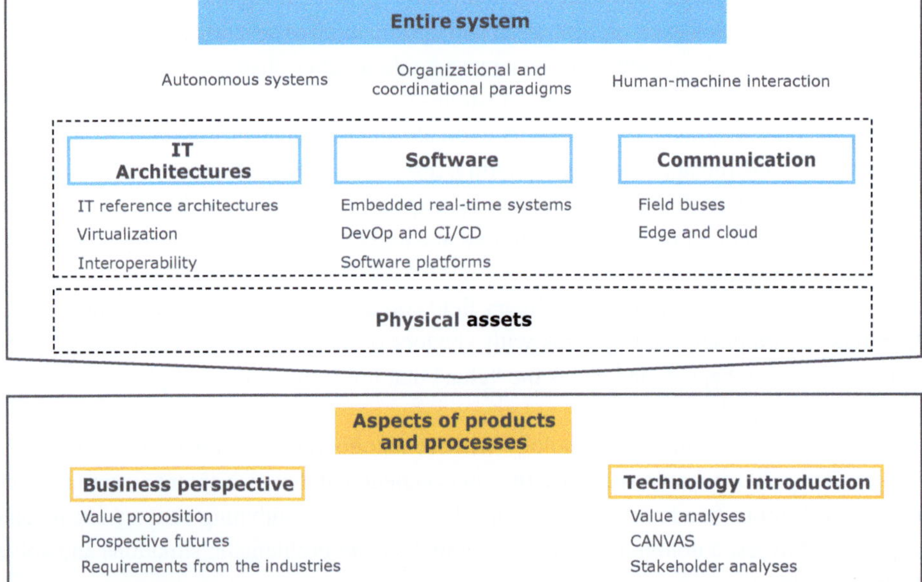

Fig. 1.2 Topology of industrial automation and information technology

are well-established and have been around for a long time while others are new and require a more profound understanding before they can be adopted.

To ensure the success of products and industrial production systems, it is crucial to consider the entire system and the targeted deployment of technologies. IT architectures, software, and industrial communication patterns utilizing a wide range of technologies are now emerging. Questions regarding reference architectures, virtualization, and interoperability are essential, along with software development for real-time systems, software platforms, and special development processes. Industrial communication—the connectivity between automated subsystems in real time—plays a vital role too. It is crucial to have knowledge of the possible solutions and use the most suitable ones.

In today's world, automation technology is increasingly being evaluated from the perspective of the introduction of a business or technology, and aspects of products and processes are considered on the basis of their value proposition and economic efficiency. It is also crucial to realize that, once the initial excitement over new technical capabilities has subsided, there will be a phase in which they have to prove themselves from a business point of view, and value analysis methods are part of the repertoire of today's engineers.

It is also crucial to pose some essential questions about the deployment of automated systems in the product creation process. This book provides readers with an understanding of the options available for successfully implementing automation methods and processes. As Fig. 1.3 depicts, this book is divided into three parts, the first of which deals with the technological basics of automation and covers relevant concepts, principles and schools of

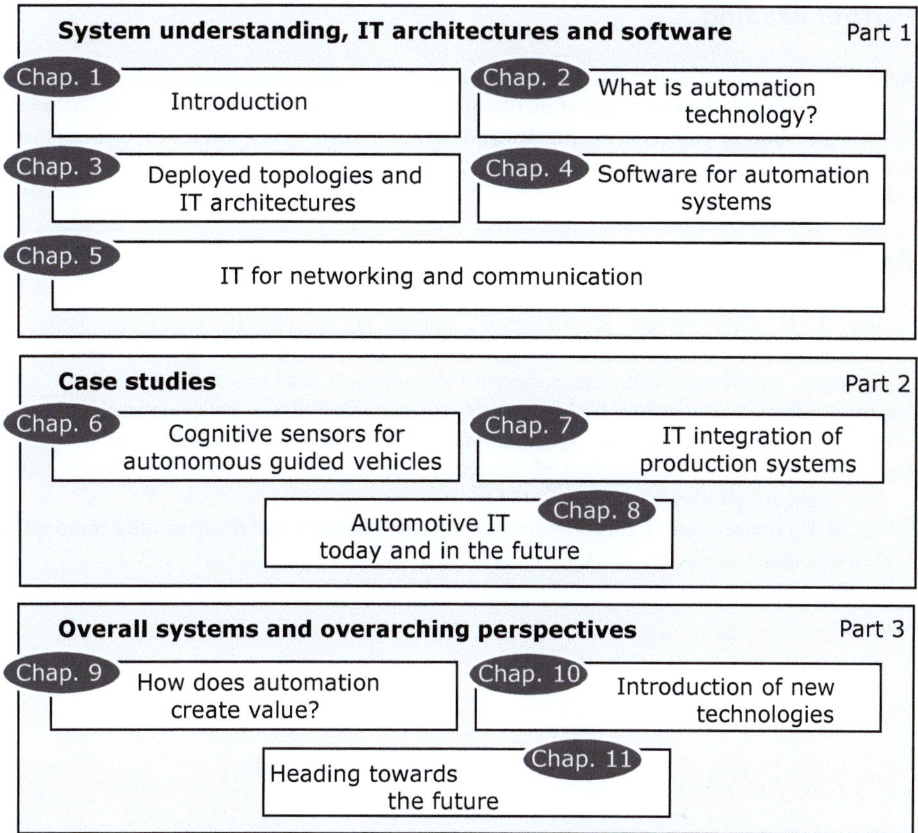

Fig. 1.3 Structure of the book

thought enabling the automation of processes and systems. The second part analyzes how and why automation technology is used today, presenting case studies from the realms of cognitive sensors, IT integration, and interoperability in industrial automation and IT for automotive purposes. Finally, the third part explains how value can be created through automation and how it can be introduced in day-to-day practice, giving readers the competence to make a systematic choice between various feasible alternatives on the basis of differing technical requirements. It explores value analysis methods and offers a comprehensive review of how emerging technologies can be assessed and implemented.

The book provides the skills necessary to understand and successfully utilize IT and software in automation technology, thus enabling effective decision-making with regard to technology selection.

Further Reading

Anderl, R.; Fleischer, J. (Ed.): **Guideline Industrie 4.0, Guiding principles for the implementation of Industrie 4.0 in small and medium-sized businesses.** VDMA and Partners, 2015

ZVEI: **Data Sharing Models in the Electro and Digital Industry.** Publisher ZVEI, German Electro and Digital Industry Association, ZVEI.org, 2022

References

Adamczyk, H.; Bettenhausen, K.; Daum, W.; Dirzus, D.; Figalist, H.; Heim, M.; Jumar, U.; Leonhardt, S.; Roos, E.; Urbas, L.; Winterhalter, C.: **Automation 2025—Hypotheses and fields of action.** (in German) VDI/VDE Society for Measurement and Control, 2015

Lauber, R.: Process Automation I – **Design and programming of process computer systems.** (in German), 2nd edition, Springer, 1989. https://doi.org/10.1007/978-3-662-09530-0

Lauber, R.; Göhner, P.: **Automation of Processes (in German) I.** 3ed Edition, Springer, 1999. https://doi.org/10.1007/978-3-642-58446-6

Porter, M. E.; Heppelmann, J. E.: **How smart, connected products are transforming competition.** Harvard Business Review, 2014

What Is Automation Technology? Basic Terms and Concepts of Automation Technology

2

Abstract

This chapter covers basic terms, topologies, and concepts related to automation technology. We shall start with an explanation of the automation technology framework.

The chapter commences with an overview of the systems view offered by cybernetics, elaborating on the key components of automated systems and exploring the relationship between human and autonomous systems. It delves into fundamental questions such as:

- What technical terminologies are used in automation, how are they defined, and what are their underlying concepts?
- What are the fundamental structures of automated systems?
- How has automation technology evolved over time, and what is its current state?
- How crucial are software, IT, and data in automation, and what are their fundamental principles?
- What is the role of humans in automation?
- What are the defining characteristics of autonomous systems, and how are they organized?

Upon concluding the chapter, readers will have a comprehensive understanding of the essential terms, topologies, and concepts of automation engineering, enabling them to provide accurate and relevant contextualization.

2.1   Basic Terms Used in Automation Technology

Automation technology is an interdisciplinary discipline that merges sensor technology, actuator technology, control technology, communication technology, real-time software, and data science, among other things. An integrated science, it strives to automate technical systems by analyzing their functions. Automation concerns the methodical design and regulation of independent technical processes, and it is utilized extensively in areas such as mechanical engineering, automotive engineering, and production engineering. According to Lauber and Göhner (1999), automation technology is defined as a technology that enables the automatic execution of tasks.

> An automated system consists of the technical system with the technical processes, the automation system, and human operators.

Today's automated systems consist of mechatronic modules, electronic systems, and an ever-increasing amount of software. This software plays a driving role in automation technology in conjunction with networked communications and information processing methods.

2.1.1   Technical System, Technical Facilities and Technical Process

How can a technical system be distinguished from a technical facility or a technical process?

Technical System

A technical system is characterized by its inputs and outputs as well as its function and structure. It comprises several subsystems that can be arranged in a hierarchical manner and all interact with each other. Each subsystem has unique characteristics, and the correlation between its input and output values are referred to as the system's function. In essence, a technical system explains how technical components, processes, and equipment are interconnected.

Figure 2.1 depicts the structure of a technical system consisting of subsystems and system elements.

In contrast to the concrete nature of the technical facilities, the abstraction of a technical system offers advantages for its description and modeling.

Technical Facility

A technical facility is a collection of equipment, devices, and machinery that work together to achieve a specific purpose. These components are concrete implementations of technical systems that have a spatial relationship and can interact with each other.

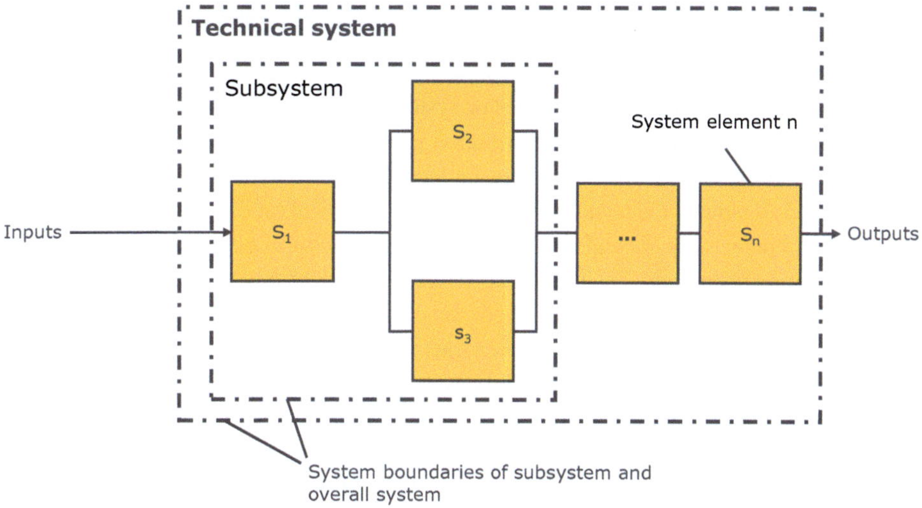

Fig. 2.1 A technical system consisting of subsystems and system elements

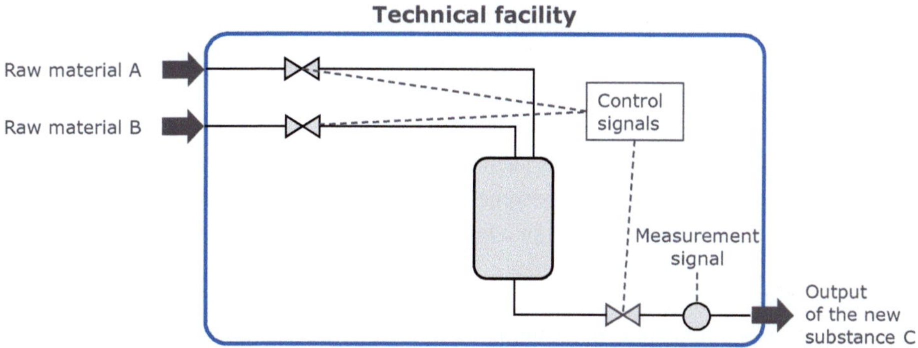

Fig. 2.2 Example of a technical facility (adapted from Lauber and Göhner, 1999; courtesy of © Springer-Verlag GmbH Germany 2024

Figure 2.2 provides an example of a technical facility comprising various components such as valves, piping, and reactors.

With the arrival of information processing, automation shifted its focus towards the abstraction of technical devices to form technical processes. This transition requires us to move from a concrete view of technical installations to a consideration of the consequences and interrelationships of activities in a process.

Technical Process

In turn, a technical process breaks a function down into a series of execution steps. According to DIN IEC 60050-351 (2014) and the VDI Glossary of Terms (2019), it is defined as:

The complete set of interacting operations in a system by which matter, energy, or information is transformed, transported, or stored.

This abstraction of technical facilities or systems in terms of function and operation is particularly useful when real-time information processing is required and the integration of distributed information poses a significant technical challenge.

Figure 2.3 illustrates the technical process of the technical facility displayed in Fig. 2.2.

Figure 2.3 depicts a technical process comprising three subprocesses—filling, reacting, and discharging chemicals. The control and state variables of this technical process can be measured and help to control the signals of the components of a technical process.

A technical process is essentially an abstract representation of a sequence of functions leading to a certain result. Technical processes can be implemented through technical systems or concrete technical facilities.

Hence, the individual aspects of the concrete realization shown in the above example can be presented abstractly in the form of descriptions of systems and processes as follows:

- The inflow of substance A and substance B and the outflow of substance C in the technical facility are presented abstractly as the input values "substance inflow A" and "substance inflow B," and the output value is termed "reactor product C".
- Real components such as valves, piping, reactors, and sensors can be represented as corresponding technical systems and described via their ability to process functions such as "filling," "reaction," and "discharge".
- Control and measurement signals such as the degree of valve opening or the measurement signal of a flow rate thus become the description of the inputs and outputs or of the command and state variables of a technical process.

There are plenty of other examples of technical processes. These include the transportation of units in intralogistics, the refining of various hydrocarbons, or the generation and distribution of energy. Another example from a totally different domain could be the plan-

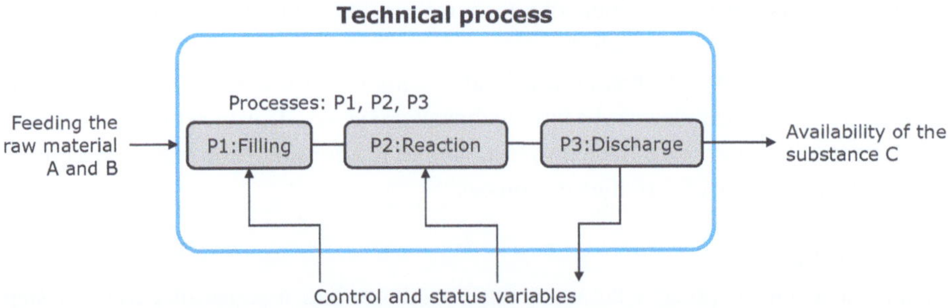

Fig. 2.3 Example of a technical process (adapted from Lauber and Göhner, 1999; courtesy of © Springer-Verlag GmbH Germany 2024. All Rights Reserved)

ning and execution of a flight or the evaluation of information, which also follow technical processes.

2.1.2   Automata and Automation Systems

An early definition of an automaton from the beginning of automation sees it as clearly separated and isolated from other systems by its functional scope. DIN IEC 60050-351 (2014) defines automata as follows:

> An automaton is a self-acting artificial system whose behavior is governed either discretely according to given decision rules or continuously in time according to certain defined relationships, while its output variables are created from its input and state variables.

In step with the advancement of information and communication technology, the traditional view of individual automata and their functional scopes is changing. Today's automation technology involves the orchestration of subsystems that work together autonomously to perform tasks. The focus is not on individual functions, but on how the functions of these subsystems are connected.

Automata are considered as the basic components of automation technology.

The definition of an automaton as a fundamental component of automation has its origin in clearly delineated technical subprocesses, each of which is processed by an automation system. An automaton is the combination of an automation system and a technical process, and it is monitored by the user.

Figure 2.4 outlines an automaton of this kind and its components: the technical system and the automation system.

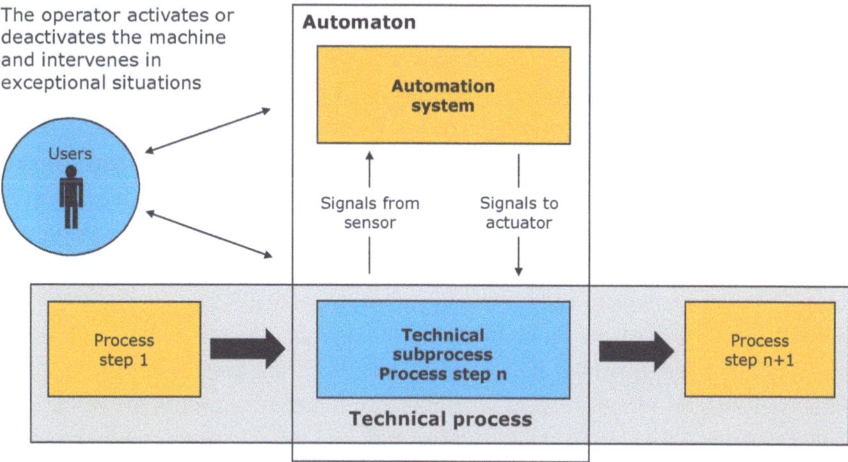

Fig. 2.4 Structure of an automaton in relation to the technical process, the automation system and the human user

In this system, the human has a special monitoring function for the automaton. The user decides on the use of the automaton in a well-defined context.

The conceptual model outlined in DIN IEC 60050-351 (2014) defines the term "degree of automation" as follows:

> The degree of automation is the proportion of automatic functions in relation to the entire set of functions of a system or facility.

The conceptual framework of automation covers automata with varying degrees of automation and corresponds with the Taylorian principles of scientific management, which are also known as Taylorism. Since the early twentieth century, this principle has had a profound influence on industrial production by atomizing processes and meticulously regulating individual workflows based on labor studies and preparatory management. Automation is then derived from these detailed tasks within individual technical processes.

Consider, for example, an industrial welding robot tasked with laying a weld according to a pre-programmed sequence. The welding process can be precisely specified and executed by either a human or a robot. This allows us to make a direct comparison by measuring the time a human requires and comparing it with the time taken by a robot. In addition, the quality of the work can be assessed. It is assumed that the mechanized welding process produces higher quality and more reproducible results, while the result of a human welder depends on individual skill. Cost comparisons can also be made. The welding machine requires investment in automation technology and incurs operating costs that compete directly with the costs of human labor and welding equipment. In addition to an economic comparison based on time, quality, and cost, aspects of flexibility should also be considered. It takes time to set up and program a machine, whereas a human welder can perform tasks immediately.

Taylor's approach has proven itself in industrial automation for many years. When applied systematically, however, it eventually leads to a spiral in which initial advances become saturated, thus limiting further improvements.

This challenge to scientific management is known as Taylor's vicious circle, in which the increased refinement of subprocesses in management requires highly specialized automation functions. According to Taylorism, all functions in the production process, such as component clamping, handling, transport, quality control etc., are considered separately and potentially performed by robots or humans. However, this complex automation is time- and labor-intensive because machines are inflexible due to complex technical dependencies, and modifying them calls for significant engineering effort.

For this reason, efforts are being made in automation to increase flexibility by implementing sensors. For example, if thermal distortion of the component due to the heat introduced during the process can be automatically compensated for in the welding robot, the robot remains limited to its dedicated function but doesn't require reprogramming by humans when the component changes due to heat input.

It is clear that efforts to deploy software in conjunction with standardized and modular systems are directed at increasing the applicability and flexibility of integrated automated systems.

2.2   The Systems Theory of Cybernetics in Automation

The systems theory of cybernetics has been a major influence on automation technology for many years. Beginning in the late 1940s, cyberneticist Norbert Wiener, theoretical biologist Ludwig von Bertalanffy, and information theorist Claude Shannon left their mark on the field. Systems theory is an overarching approach to thinking that follows the system and its properties. Since then, systems theory has undergone intensive development, and now it has many points of contact with information, control, and systems engineering as well as operations research.

2.2.1   The Feedback Control Loop and its Elements

It is essential to have a comprehensive understanding of the fundamental components of automated systems and the way in which they are categorized. The technical automation process involves numerous components such as actuators, sensors, and a control system.

Figure 2.5 illustrates the functional blocks present in a closed-loop control system. The example shown here is a closed-loop control that acquires information from the technical process through sensors, processes this information in a controller, and then influences various functions of the technical process by means of an actuator. Sensors and actuators are the interfaces of the automation system. They process the sensor signals or control signals input into or output from the controller. A so-called reference signal determines the output variables of the technical process. The control loop is closed, so the output vari-

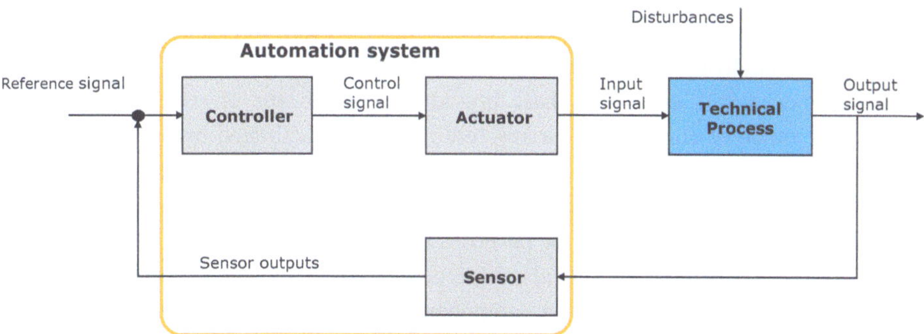

Fig. 2.5 A feedback loop control and its elements

ables go on corresponding with the reference signals even in the presence of unknown disturbances or deviations from the expected behavior of the technical process.

This structure does not require us to have a precise understanding of the relationships between input and output variables. The output signals are measured by a sensor. This allows the control mechanism to adjust the output variables via feedback to minimize the difference between the reference signal and the actual outputs. The feedback allows the sensors to detect the influence of disturbance quantities so that they can be compensated for whenever necessary.

The sensors and actuators must be positioned in close proximity to the technical process in which information is sensed and actuators carry out physical interventions.

In the past, for technical reasons, the open- or closed-loop control was located in close proximity to the sensors and actuators. Today, however, information technology networking has made this proximity less necessary since data and information processing can now be virtualized and executed at different locations.

2.2.2   Actuators

Actuators are devices used to transform control signals into mechanical work or other physical process variables.

For example, the actuator may use additional energy to produce a velocity, a force or motion along a certain path. This allows the physical variables of the process to be influenced. It also means that, in addition to mechanical work, actuators can generate other physical variables such as sound or light or affect the temperature. Actuators use different forms of energy to perform the physical activity: depending on the actuator, electromechanical, fluidic, thermal, or even chemical energy can be used.

Due to their close proximity to the technical process, actuators play a central role and are the subject of independent disciplines dealing with the corresponding physical and chemical contexts.

Currently, a wide range of commercial products are tailored to automation applications. The operating principles and the relevant structures are usually very task-specific, but they can generally be divided into three distinct areas.

Electromechanical actuators: These provide various means of controlling electrical drives. The mode of control plays a key role too and is facilitated by power electronics such as converters and gears. These actuators have been undergoing intensive development for decades, leading to numerous designs and configurations. For example, servo motors equipped with sensors for the purposes of precise positioning and drive control are now available. These servo motors also have control electronics and software which turn them into self-contained automation components. Electromechanical systems are widely used in industry, and numerous manufacturers of drive systems tailor their products to the specific needs of various applications. These include a wide range of DC, asynchronous, and synchronous machines aa well as linear and stepper motors.

Fluid systems: Fluid systems use either compressed air or a fluid (usually oil). There are motors for rotary motion and cylinders for linear motion, while various valves, compressors or pumps ensure that the pressure is distributed. Basically, hydraulic systems allow high forces to be transmitted with the help of fluids. However, the high pressures can cause oil to contaminate the environment. For this reason, hydraulic systems are limited to applications in which environmental impacts of this kind can be controlled. Compressed air is also widely used in industry to drive actuators. These systems are environmentally friendly because the medium is air and they have lower requirements with regard to piping. Manufacturers in this field also specialize in specific application areas and offer dedicated products designed for the purposes of automating production systems in particular.

The great variety of physical effects means that thermal and chemical systems contain many highly specialized actuators. Manufacturers offer specialized solutions such as controlled micron motion based on piezo actuators, memory metals to generate motion, and other effects such as magnetostriction or electrolytic pressure.

2.2.3   Sensors

Sensors play a vital role in automation systems by serving as their sensory organs. They are used to detect the state of elements within a technical process or environment and are also called detectors or transducers because they convert physical or chemical quantities into signals, process them, and communicate the data collected.

A sensor requires a setup for calibration or adjustment. So-called "intelligent" or "cognitive" sensors have the ability to self-adapt to allow automatic error correction or self-diagnosis of their own state.

Sensors attempt to imitate the five human senses by perceiving a wide range of properties and characteristics, as shown in Table 2.1. A vast range of sensors involving complex technical approaches and applications are now available to automation technology.

There are various effects that can be used as measuring principles for sensors, leading to a wide range of variants for determining the measurable variable. Depending on the

Table 2.1 Sensors—the sensory organs of automation systems

Sense	Perception of:	Examples for sensors
Visual (vision)	Light and contours, scenarios	Camera, optical measurement technology
Auditory (hearing)	Sound	Microphone, ultrasound
Olfactory (smell)	Fragrances	Analyzer
Gustatory (taste)	Flavor/ingredients	Chromatograph
Tactile (touch)	Forces, moments, shapes, positions or heat	Probe, strain gauge, thermometer

application and the technical possibilities, a combination of measuring principles can be useful too.

In the field of automation, a wide spectrum of sensors is currently available. The technical approaches and applications are extremely diverse. A broad range of effects can be used as measuring principles for these sensors. Depending on the specific application and the technological capability, this results in various approaches to measuring the relevant quantity, and it can be of advantage to combine several sensing principles.

Physical quantities such as pressure and force can be measured using various effects. Mechanical spring-damper systems, for example, allow forces or pressures to be derived from mechanical displacement. Numerous other approaches are suitable for different measurement ranges, such as the piezoelectric effect, which generates an electrical voltage as a measurable quantity. Temperatures can be measured using the thermoelectric effect or semiconductor-based methods involving voltage, resistance, or current changes. Light intensities can be assessed using photodiodes, which measure a current. Distances and lengths can be determined by pulses or voltages generated by electromechanical or optoelectronic systems.

Sensors are the sense organs of automation, and they play a central role in many applications. History has shown that the availability of novel measurement techniques can revolutionize certain aspects of automation. A wide range of sensor systems has evolved to meet the diverse requirements of technical processes. In the event of sensor system failure, a variety of alternative products can be selected. But how should one go about selecting a sensor system? The first question to ask is which physical measurement principle is used to detect a particular variable. For example, length can be measured by inductive, capacitive, magnetic, or optical means. It becomes critical to consider functional requirements, system requirements, and various ancillary functions. The choice of measurement principle is a fundamental decision that affects not only measurement accuracy and tolerance but also cost-effectiveness. However, selecting an appropriate measurement principle or commercial sensor is often challenging and requires experimentation to demonstrate feasibility and ensure that the practical requirements are met.

Modern sensors are more than just data collectors that record temperature trends in time series. They offer a wealth of additional functions such as processing measurements, automatically correcting measurement errors, and providing support for configuration and maintenance.

This offers the potential to improve data acquisition, provide specialized support for configuration and operation, and offer additional functionality through specialized programs:

- Multiple measurement principles can be harmonized to merge the data, enabling data fusion to improve the quality of measurements or conduct validity checks for error detection.
- A system-integrated control mechanism can organize the setup process. This is facilitated by a configuration database that supports the configuration and adjustment of measurement technology during operation by relying on established parameter sets.

- Operational data can also be collected through sensors that allow self-calibration or self-monitoring through specialized programs.
- Human-machine communication, for example, can be structured using applications in order to facilitate operation by untrained personnel. Personnel should be assisted in selecting system configurations during commissioning or maintenance.

While these enhancements significantly increase the complexity of a sensor system, end users perceive "intelligent sensor systems" as facilitators due to their ease of use and the operational support they provide.

Also, the interaction between sensor networks is now becoming increasingly critical, emphasizing the growing importance of sensor data fusion. This includes the processing of different data structures as well as their correlation and evaluation. Today's sensor systems often have significant computing power that greatly exceeds that of the controller.

The next step towards advanced sensor technology is achieved using cognitive sensors. This involves advanced signal processing moving towards perception, i.e. the first stage of awareness, or the integrated interpretation of data. Functionalities of this kind depend on the ability to process different measurements or combine and jointly evaluate them with the help of sensor fusion.

As a result, sensors are poised to become even more significant as autonomous subcomponents in the future.

2.2.4 Types of Control Systems

Control is at the center of an automated system. Despite concerted efforts to harmonize terminology, this term is often used imprecisely.

However, a fundamental distinction can be made between "open-loop" and "closed-loop" control, the last of which includes a feedback loop.

There are two types of open-loop control: control without feedback and control that detects and considers at least some disturbances and takes them into consideration for feedforward control.

Feedforward control is shown in Fig. 2.6. The relationships between input and output variables have to be well known at the time of design. On the basis of this knowledge, the control signals of the control can be implemented to drive the technical process towards the desired output signals. If the disturbances are known, the control system can be designed to compensate for them. However, when unmeasurable or unknown disturbances affect the system, their effects propagate to the output without compensation. Since this is quasi-anticipatory control without feedback, deviations in the behavior of the technical process can lead to potential difficulties as the control may operate inaccurately.

In contrast, closed-loop control refers to a closed-loop system with feedback of the output variables (feedback control, see control loop in Fig. 2.5). In modern automation systems, both methods are often used together to provide enhanced functionality. For

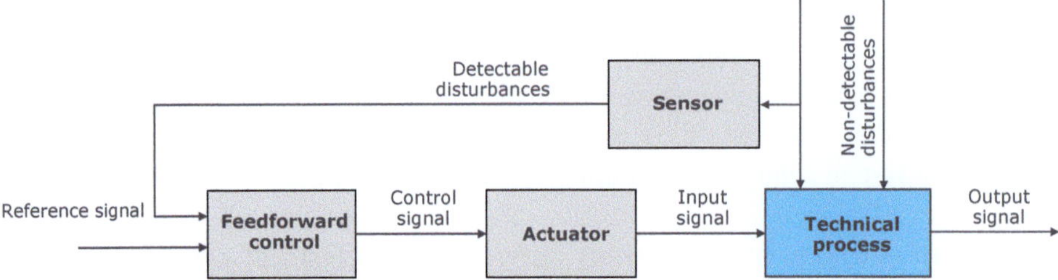

Fig. 2.6 Feedforward control with detection of disturbances by a sensor

example, in a heating thermostat, closed-loop control is used to regulate the inlet valve of the radiator on the basis of temperature sensor readings. The controller aims to keep the temperature as close as possible to the setpoint selected.

However, a system of this kind can become destabilized by unexpected changes. For example, when windows are opened for short periods of ventilation, a closed-loop system would attempt to maintain the room temperature, causing energy to be wasted. However, when the temperature sensor signal is used for open-loop control, it can detect the sudden drop in temperature, prompt the system to disable control temporarily until the brief ventilation ends, and then resume temperature control.

2.3   The History and Advancement of Automation Technology

In recent decades, automation technology has experienced numerous innovations due to developments in information and communication technology as well as software. Automation technology has always focused on the technical process, its optimization and its functional expansion.

Table 2.2 shows these stages and refers in particular to the design of the automatic control.

Automation technology has significantly altered the manufacturing industry and its competitive landscape on multiple occasions. In the mid-1960s, mechanical and electronic automation was introduced for individual functions such as welding and mechanical processing. The first ever programmable logic controllers (PLCs) came into use around 1970 with the advent of transistors and microprocessors. This paved the way for replacing hard-wired logic control by software-based programming.

Advancements in microelectronics and communication technology in the 1980s and 1990s led to the increased use of microcomputers, distributed controllers, real-time control and process monitoring as well as the standardization of fieldbus systems. Since 2000, the development of integrated "embedded systems" has increased significantly. Systems of this kind use higher programming languages and object-oriented software underpinned by a unified approach.

Table 2.2 Stages in the development of automation technology in recent decades

1960s/1970s	1980s/1990s	2000s/2020s
Individual automated activities based on mechanics and electrics/ electronics	IT-driven change based on microelectronics and software	Information and communication technology become an integral part of the product.
Trend: Logic controllers, encoders, and drive technology First programmable logic control (PLC)	Trend: Microprocessors and computers, real-time programming languages, structured design, fieldbus systems, and motion control.	Trend: Creation, operation, and networking of software systems and use of cloud technologies. Embedded systems with connectivity and object-oriented software systems

Today, networked communication and information processing methods are well established in the field of automation technology. Since 2010, increased digitalization has been having a strong impact on automation technology. Initially, work focused on systems engineering with the aim of better organizing the emerging diversity of embedded processors and sensors and improving the connectivity of systems. Since around 2020, though, software has become the leading variable in automation technology.

Emerging technology platforms that map basic functionalities are increasingly being used as industry standards. Cloud computing has shifted the focus to the field of virtualization, thus breaking the link between physical location and processing via information technology. The shift is from location-based approaches to signal flow- or data-driven architectures able to process large amounts of data. The removal of the physical constraints on the location and position of electronic systems paves the way for the design of new software systems.

These technical developments are changing the structures of automation systems. There is a growing desire to network systems with ubiquitous subsystems that exhibit autonomous capabilities and are closely integrated into our everyday lives. Our attention is now shifting from cybernetic sets of activities to autonomous systems. Figure 2.7 illustrates this extension and the shift in terminology it entails.

A cybernetic automation system follows a three-step process of sensing, planning, and acting. In a process of this kind, algorithms perform automatic control with the help of measured variables. For example, the speed of a vehicle can be controlled using a closed-loop system which is based on known relationships and includes a tachometer, a controller, and a power unit.

As the system evolves, it will develop cognitive capabilities that will extend beyond the three-pronged approach described here. In autonomous systems of the future, the inner circle will be surrounded by an outer one involving perceiving, understanding, and problem-solving.

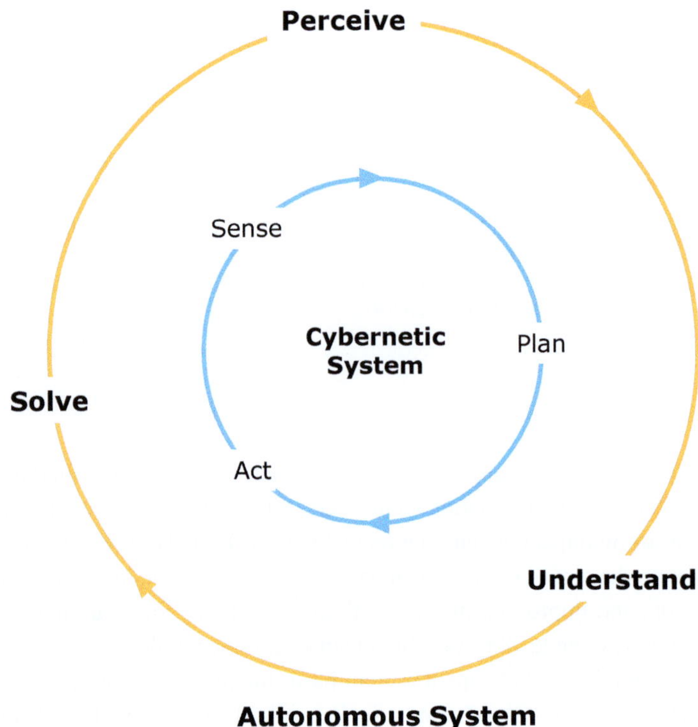

Fig. 2.7 From cybernetic systems to autonomous systems

For instance, an autonomous vehicle would use sensor systems to perceive its surroundings. The information provided by the sensors would then be used for comprehensive pattern recognition, thus allowing complex alternatives in terms of action. This process requires the system to possess specific "skills" enabling it to perform a task independently on the basis of knowledge and expertise.

In step with the evolution of algorithms, simple automation systems with well-defined functions will increasingly be replaced by the more complex cognitive skills of autonomous systems.

2.4   The Impact of Software, Information Technology, and Data on Automation Technology

Nowadays, software is extensively utilized for automation purposes. In addition, the advancement of information and communication technology (ICT) has facilitated the interconnection of systems, generating vast quantities of data that can be utilized for automation. The current developments in the field have gone far beyond this, utilizing layered data and information enabling systems to act autonomously as clearly recognised by Holdren et al. (2010).

2.4.1   Software-Intensive Automation Systems

The utilization of network communication software has paved the way for various innovative and value-driven developments. As a result, software systems have become indispensable elements of today's advancements in automation technology. This begs the question as to the primary areas in which automation software will be put to use.

A software system depends on a set of software subsystems—also known as software components—which communicate with each other and run on one or more computers in order to govern a mechatronic system, for example.

Software has become increasingly omnipresent in our daily lives. Software applications vary widely, from smartphone applications for personal use to software solutions for industrial product management in companies and from medicine, transportation, and aviation to energy generation. The development of software systems has escalated remarkably in recent years, and it is remarkable to see how large and complex the code used in automotive systems has become.

This trend is not surprising given the growing complexity and interconnectivity. In the future, the technical infrastructure will rely more on software systems and include a constant proportion of mechanical components. Software systems will interact with physical devices, processing information and making decisions based on networked information and decision-making processes. In this way, mechanics, software, communication technology, and information technology will come together and serve as a basis for future developments.

In step with the shift from mechanical systems equipped with electronics and dedicated software to software-based systems allowing extensive networking, the use of software in industrial automation is expected to surge.

Surveys conducted by management consultancies reveal that the traditional technological sector of automation is leaving hardware behind and shifting its focus to software and services. Just two decades ago, two-thirds of the sales revenue of the German technology sector came from hardware, but now hardware only represents half of the total figure. At the same time, the European technology market has more than doubled within two decades.

2.4.2   Software Development Processes in Automation

The utilization of more and more software is transforming the design process and changing automation technology, resulting in a faster development process. In the past, prevailing conditions meant that advancements in mechanical engineering had to follow a structured approach divided into different phases and spanning several years. Today, however, developers enjoy more flexibility thanks to the availability of machine system components in module form.

Software development, on the other hand, requires an entirely different approach due to its intangible nature. It can be altered through updates, even during use, thus creating new possibilities for agile development. However, this presents challenges for system development too.

As shown in Fig. 2.8, the development process has evolved from a sequential approach to a more flexible and agile methodology.

It is becoming more and more common for systems to undergo changes during deployment, such as during maintenance or rebuilding phases, or even after a system is restarted. For example, when a car is taken to the garage or a dishwasher is repaired, a new software update is typically installed to improve the characteristics of the product or system. We have all become accustomed to consumers connected to the Internet receiving updates without our knowledge.

The concept of DevOps, which stands for "Development" and "Operations," suggests that a system's functionalities can be continuously developed while it is in operation. This means that software updates can be made even after delivery, allowing for continuous improvement and adaptation to customer needs. However, there are also risks associated with DevOps as the ability to modify automation systems can lead to security issues.

The CI/CD process is often used in the context of DevOps. It stands for continuous integration/continuous delivery and continuous deployment. It is an automated process used by development teams to generate and deliver software updates to systems in the field frequently and at short intervals.

The CI/CD concept involves three phases:

1. Continuous integration allows code changes to be managed in a common tree. The software goes through an automated testing and release process that reveals existing conflicts in the code. This mechanism favors faster application development, bug

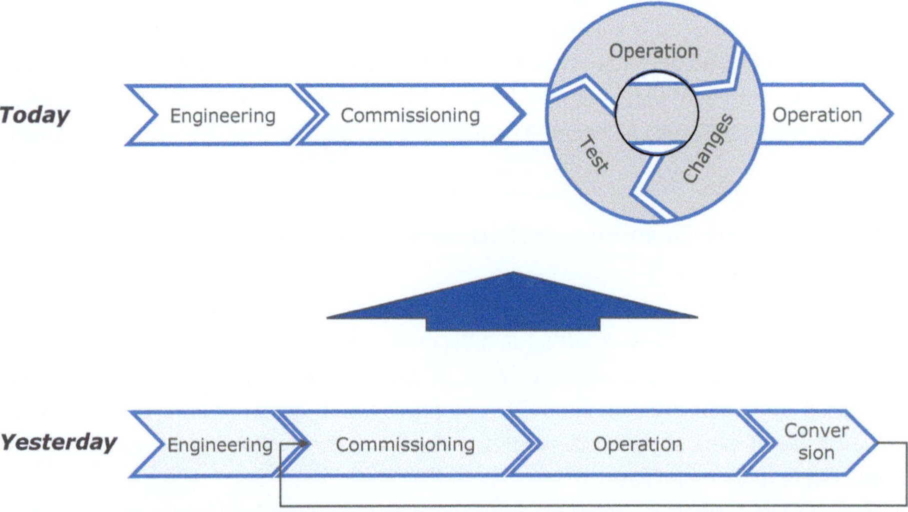

Fig. 2.8 Change in the development process to adaptation during operations

detection, and incorporation of customer feedback, leading to improved software quality.

2. The next phase of the software development feedback process is known as continuous delivery. This builds on continuous integration and is used for automated code validation and functional testing so that any bugs can be detected and fixed. This stage is also used to release code automatically so that updates can be rolled out at any time.

3. Finally, the automated rollout of updates known as continuous deployment takes place. Software modules are automatically released and rolled out here. The correct use of the CI/CD concept is particularly relevant for systems that must be operated and kept up-to-date in the long term without the necessity of accepting downtime due to updates.

While over-the-air updates of operating systems in mobile phones and laptops are standard practice, they are not yet widespread in technical systems such as those under consideration here. Software updates can enable novel system capabilities that were not foreseeable at the time of the original development. As a result, new capabilities in software development raise many questions regarding the functional safety, quality, and sustainability of automated systems.

2.4.3   Cyber-Physical Automation Systems

The use of software and technology services is becoming increasingly prevalent due to rising levels of connectivity, artificial intelligence (analytics, machine learning), and the emergence of innovative business models for service delivery.

The evolution of technology towards the ubiquitous presence of information-processing networked systems has been anticipated for years. As early as 1999, a visionary named Marc Weiser wrote an article entitled "The Computer for the 21st Century," where he predicted that the then-dominant personal computer would disappear as a stand-alone device and be replaced by networked objects instead. It was in this context that he coined the term "ubiquitous computing," which refers to the omnipresence of computers.

Technologist Kevin Ashton is credited with coining the term "Internet of Things" (IoT) in 1999. The IoT allows uniquely identifiable physical objects such as those occurring in logistics to be linked to a virtual representation in an Internet-like structure.

In addition to the growth of software, the automation technology of today is characterized by a situation in which an automation system generally consists of many computers networked with each other using communication technology and joining forces to create a virtual representation of the real world.

It is expected that future automation systems will consist largely of software, be highly networked, and will process extensive amounts of data. This will lead to the prevalence of information systems that process networked data in a way reminiscent of the availability of electricity today.

It was in this context that Edward Lee popularized the term "cyber-physical systems" at a workshop sponsored by the National Science Foundation in 2006.

According to the VDI/VDE-GMA glossary,

> a cyber-physical system (CPS) is defined as a system which links real (physical) objects and processes with information-processing (virtual) objects and processes via open and in some cases global information networks which are connected at all times.

Since then, there has been a lot of activity around cyber-physical systems. This is in step with the emergence of a world of information technology that complements the physical world and simplifies interactions with it.

In 2014, the term "hybrid reality" was coined by Eric Schmidt, the former executive chairman of Google, to describe an interaction between a world of physical systems and a new realm of digital information. It raises questions about the implications of a future in which the majority of the world's population is online and information technology is always available.

The U.S. President's Council of Advisors on Science and Technology (2010) addressed this issue in a letter to the U.S. President entitled "Designing a Digital Future: Federally Funded Research and Development in Networking and Information Technology". This outlined a digital future and underscored the societal importance of the impending digitalization.

In Germany in 2010, the topic of cyber-physical systems was taken up by Acatech, the German Academy of Science and Engineering. In 2011, the term "Industrie 4.0" was first used at the Hannover Messe as a recommendation for the upcoming fourth industrial revolution based on information and communication technology in industrial production.

Cyber-physical systems are characterized by distributed control, which enables the targeted control of individual components of the industrial automation systems. Information networks form the basis for linking specific technical subprocesses. Data and information are not stored in one place, but are distributed and accessible from anywhere. This decentralized data access is often referred to as "cloud computing", emphasizing that the critical factor is not the physical location of the data but the existence of a comprehensive system of interconnected computers.

Within this framework, the concept of the digital twin emerges, meaning a digital replica existing in parallel with the physical system. Different systems are interconnected through networks, allowing data and services to be used in different locations and forming an information space parallel to the physical world.

Cyber-physical systems enable the following three paradigms:

1. Distributed control using semi-autonomous systems: Subsystems can act autonomously based on the availability of distributed information and perform activities in coordination with other subsystems. Individual subprocesses can be controlled sepa-

rately but coordinated with the help of an overarching control which allows decentralized technical processes to be organized or adapted to changing conditions.

2. Novel information spaces based on infrastructures for managing large data sets: Their technical implementation occurs in a cloud and runs on the computers of major IT providers around the world. The accumulation of data means that access to and use of these data sets remains open and subject to interpretation within the context of national laws and societal policies.

3. Runtime evolution of software systems: Software reconfigurations take place not only during development but also in the field during use. In the future, technical systems are expected to evolve constantly during operation. Whereas systems were structured and uniquely configured at the time of development in the past, it is expected that interactions between system components that are not fully predictable during the development phase will become widespread.

All three aspects allow for adaptation to environmental conditions. This will also lead to the emergence of a world of information parallel to the physical world which will in turn affect the latter.

Informed decisions about changes in the technical process can be made and implemented on the basis of precise knowledge of system states during operation, comparison capabilities, and intervention options. An important future consideration is the way in which software can be used to make changes such as process adjustments or modifications. However, it must also be possible to make much more complex adjustments to mechanical and electronic/electrical hardware.

2.4.4   The Digital Twin

Recent trends in research and development highlight the increasing popularity of digital twin technology across multiple industries. Essentially, a digital twin is a virtual replica of a physical system or component created using software and data. This enables it to accurately imitate the behavior and characteristics of its physical counterpart in real time. The digital twin is a powerful asset that offers a comprehensive understanding of the physical system, thereby facilitating better analysis, prediction, and optimization of its performance.

According to Ashtari et al. (2019), a digital twin is defined as follows:

> A digital twin is a general service architecture comprising multiple models and software services which interact with the real asset. A digital twin consisting of models and software services is created in unison with the physical system and its assets.

Previous definitions of the digital twin characterized it as an integrated multiphysical simulation of a system. The concept has been used as a software model for developing and

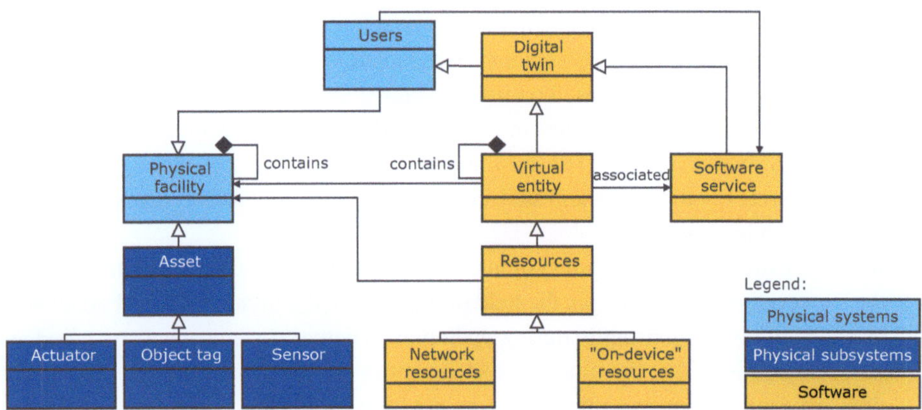

Fig. 2.9 Class diagram with physical facilities, user, and digital twin (adapted from Carrez 2013)

testing various configurations. However, simulation is only one of many aspects of the digital twin. Digital twins are now being created in particular for the purposes of virtual commissioning and field data acquisition. Digital twins are based on a functional digital model which captures the current state and behavior of the technical system. The digital twin emphasizes the virtual space and enables control, regulation, and monitoring even within large and spatially distributed technical subsystems—even if they undergo complex changes and different subprocesses merge within changing system constellations.

A digital twin describes significant characteristics and features of the technical facility, focusing on a selected context such as system geometry, electrical design, software, or specific operational data. It captures the properties, conditions, and behavior of the real system through models and data, with the virtual element representing its physical counterpart.

Figure 2.9 shows a UML class diagram that distinguishes between the digital twin, the physical facility, and the user. The digital twin represents the so-called base class for the virtual entity and the software services associated with the data coming from the physical facility.

The UML class diagram also illustrates the relationship between the physical asset, the user, and the software.

It is relevant to note that the physical facility and the virtual entity have a reflexive compositional relationship, meaning that they can be formed from different instances to ultimately form a whole.

2.5   The Role of Humans in Automation Technology

Despite advances in automation, humans still have key responsibilities that require responsible fulfillment:

- Designing and developing automation systems, including determining what functions should be automated or what capabilities an autonomous system should have.
- Commissioning of the automated system and its specific configuration as well as adapting to the conditions of the technical process.
- Operation and monitoring of the automated system, including the responsibility for switching it on and off and ensuring its operation.
- Maintenance and upkeep, in which humans ensure long-term operation and minimize downtime.

Developing appropriate human-machine interaction is a challenge. Depending on the use of automation systems tailored to specific applications, numerous tasks must be performed within the framework of human-machine interaction.

The operation or use of automation systems entails a number of different requirements. The two examples shown in Fig. 2.10—the human-machine interface of a control room and a servo motor—illustrate this diversity.

It is apparent that a control room demands a high level of monitoring, observation, information presentation, and support for control intervention. In contrast, the human-machine interface of a servo motor should support configuration during commissioning and diagnostics. This example illustrates how there are different requirements depending on the application.

User requirements for human-machine communication are also varied. The operation can take place directly at the device, as in stationary systems, or remotely, as in remote

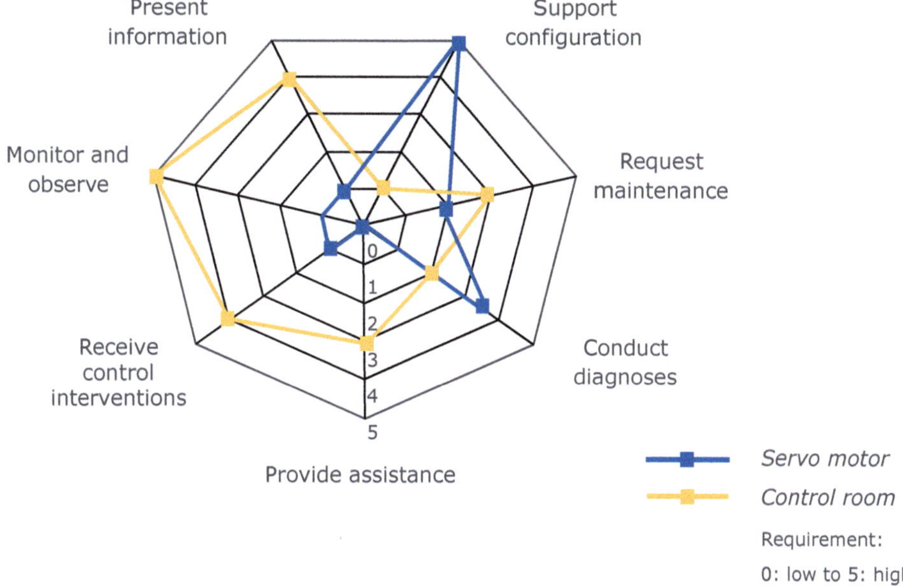

Fig. 2.10 Comparison of aspects of human-machine interaction in different automation systems

maintenance. Individuals operating the automation system have varying levels of expertise and intellectual capacity. For example, trained operators should be able to start or stop the automation system. However, extensive commissioning or maintenance tasks require skilled personnel. In addition, users come from different cultural backgrounds, speak different languages, and assign different meanings to symbols. The human-machine communication system must account for these variations.

In addition, the automation system should provide protection against user error. Automation systems, especially those used in production, are safety-critical and must be designed to prevent accidents resulting from inadvertent interactions. However, many processes cannot be fully automated. Complex tasks lead to the automation of subtasks initiated by humans before the system performs them automatically. An example of this is automated driving, in which humans perceive the support of the automation system as useful assistance allowing them to better focus on their tasks.

From a system design perspective, there are many ways to structure human-machine interfaces. Figure 2.11 provides an overview of input and output methods in automation systems and their components. In automation, many technologies have proven to meet high reliability requirements even under harsh environmental conditions (e.g. noise, dust, humidity, etc.).

For example, robust inputs can be provided in the form of buttons, encoders, mouse devices, etc., allowing tactile interaction with the system. Outputs can be LEDs, LCDs, monitors, or projections. Remote access such as that provided by means of applications is becoming increasingly important. Emerging technologies such as voice recognition, handwriting input or the use of special projection glasses are still in the research phase and are only feasible for certain specific applications.

Human-machine communication places high demands on software development environments too. The development of software systems for automation purposes requires comprehensive development platforms for system design that enable the exchange of information between implementations and its reuse.

	Electrical/ electronic components	PC-based systems	Mobile systems (Smart Pads)	Further technologies
Output	• LCD displays, LED	• Monitors, projections, wide screens • Graphical user interface of operating systems		• Special glasses
		• Web interface	• Multitouch for navigating and zooming • Apps	• Language
Input	• Buttons, rotary encoders, gyro sensors	• Mouse, touchscreen, tablets		• Object tracking • Recognition of handwriting

Fig. 2.11 Input/output capabilities for the realization of the human-machine interface

As automation technology advances, the human-machine interface becomes more and more important. In the exchange of information using control elements, the focus is on the type of execution. The spectrum ranges from very simple approaches to mobile input devices or sophisticated specialized systems such as data glasses.

In the future, automation systems will intrude more deeply into the daily lives of users through networking and the autonomous assumption of functions, thus increasing the importance of human-machine communication. The socio-technical environment—i.e. the interaction between the social environment and the technical system—is in a state of change. The integration of automation systems into our social system is increasing constantly. The availability of advanced human-like communication capabilities such as language, semantic definition, symbol recognition, and learning becomes crucial in this context.

Social interactions are becoming increasingly important, especially when exchanging information or negotiating with other systems or system users operating in a specific environmental context. For this reason, the focus is not only on the decisions of an individual user regarding direct specifications for the automation system, but also on many factors that the automation system can process in an appropriate manner. The question as to how to integrate automation systems such as health care robots or autonomous vehicles will become crucial. But how and at what levels does human-machine communication take place, and how deeply are they integrated into our social systems?

It is foreseeable that automation systems will require capabilities that will enable them to respond with empathy in socio-technical contexts. This requires a comprehensive interpretation of interactions in social systems as well as a proper understanding of emotions such as joy, enthusiasm, anger, frustration, or panic.

2.6 Autonomous Systems

Autonomous systems are designed to solve complex tasks independently and respond to unpredictable events. Unlike classic automated systems, autonomous systems can interact with operators and guide actions of their own accord. These systems are capable of performing tasks that are typically carried out by humans, and they can operate in the air, on the water, and on land.

Sifakis and Harel (2022) describe autonomy as follows:

Autonomy is the capability [of a system or agent] to achieve a set of coordinated goals without human intervention by adapting to changes in the environment.

According to Müller et al. (2021), an autonomous system is defined as follows:

An industrial autonomous system is a delimited technical system which, systematically and without external intervention, achieves its set objectives despite uncertain environmental conditions. [...] Two further characteristics are used to distinguish intelligent industrial automation systems from industrial autonomous systems: self-governance and independence.

It is anticipated that future autonomous systems will be capable of planning necessary steps situationally and within a defined range of action as if limited by guard rails.

An autonomous system performs tasks, recognizes relationships, and takes control. It can generate insights from data and plan actions.

The focus of autonomous systems is therefore on complex perceptual processes based on extensive information gathering and processing that triggers extensive actions. This involves extracting partial information and processing it into overall impressions that drive actions. Complex information processing of this kind relies on the perception and recognition of the environment (cognition) in conjunction with the derivation of appropriate actions.

The typical main components of an autonomous system are shown in Fig. 2.12. An autonomous system or agent has different main components than an automaton (as shown in Fig. 2.4).

Even in an autonomous system, process interventions are performed by actuators and process data are captured by sensors. However, the system's actions are generated by a goal-based "decision-making" component that uses both "information and knowledge processing", "perception", and "reflection" for "goal management" and "action planning".

The main components have the following roles:

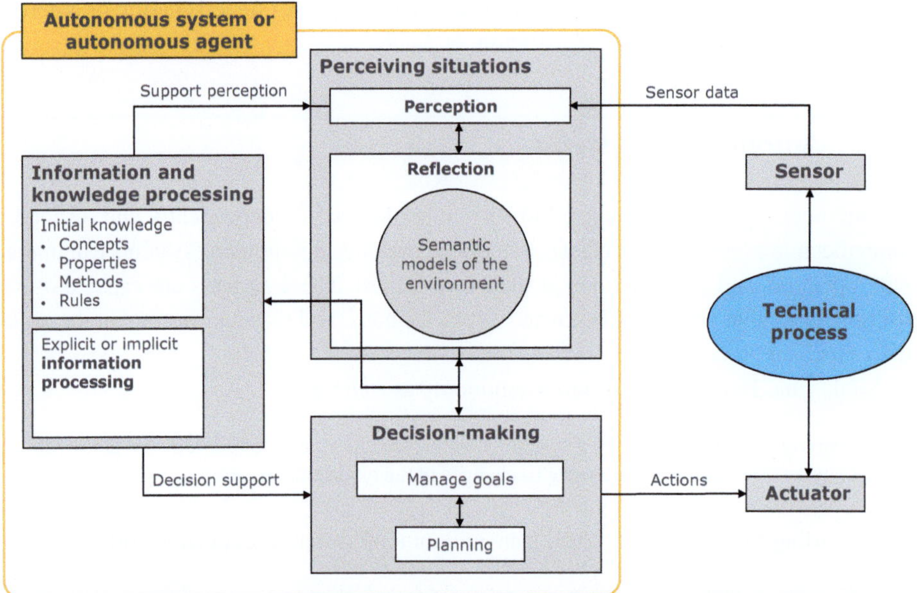

Fig. 2.12 Autonomous system architecture and its key components (from Sifakis and Harel 2022; courtesy of © ACM—Association for Computing Machinery 2024. All Rights Reserved)

- "Perception" interprets sensor data, resolves ambiguities, and gleans relevant information from complex sensor data.
- "Reflection" is responsible for building and updating a realistic runtime model of the environment.
- Goal management selects the most appropriate goals for a given configuration of the environmental model.
- "Planning" determines what is necessary to achieve the selected goals and enables the actuators to perform actions that influence the environment.
- "Processing of information and knowledge" involves the ability to supplement initial knowledge by learning new relationships. "Reflection" allows goal management and the dynamic feeding of the planning process with information to support decision-making.

In this way, automated systems or autonomous agents acquire the ability to adapt to new situations and plan actions—something which only humans were capable of up to now.

This highlights the important role of artificial intelligence (AI) and machine learning (ML) in automating processes and providing more support to humans. While progress has been made in recognizing complex patterns and deriving actions, there are still many unanswered questions with regard to deduction and inference, meaning that, for the foreseeable future, autonomous systems are likely to remain limited to specific use cases such as selected situations in autonomous driving or flying.

However, the transfer of control to an autonomous system, i.e., the assignment of responsibility for activities, raises not only numerous technical challenges but brings up legal questions too. In critical situations, decisions are no longer made by humans but by the system itself instead.

Nevertheless, many of today's use cases are fine-tuned by humans. This allows the autonomous system to recognize its limitations and prompt human intervention.

The first implementations are currently emerging in autonomous driving, where vehicles can fully automate their travel from point A to point B. Even in the case of disruptions or unforeseen influences, these vehicles can identify and implement autonomous alternative actions. Further advances in autonomy which allow complex situations to be managed even in the face of unforeseen events are critical for industrial automation in factories, plants or other facilities.

Figure 2.13 shows the five stages of autonomous system behavior. These five stages are well-established in the automotive field and were defined by SAE International (formerly the Society of Automotive Engineers) for automated and autonomous driving. The stages can be applied to a wide range of engineering systems.

In the field of autonomous driving, systems classified as Level 3 "Highly Automated" are already on the market. However, the path to Level 4 "Fully Automated" or Level 5 "Autonomous" remains unclear. Research prototypes enabling Level 4 (fully automated driving) are available, but humans still need to intervene, for example through teleoperation, to resolve exceptional situations. Today, widespread use of Level 4 or 5 autonomous

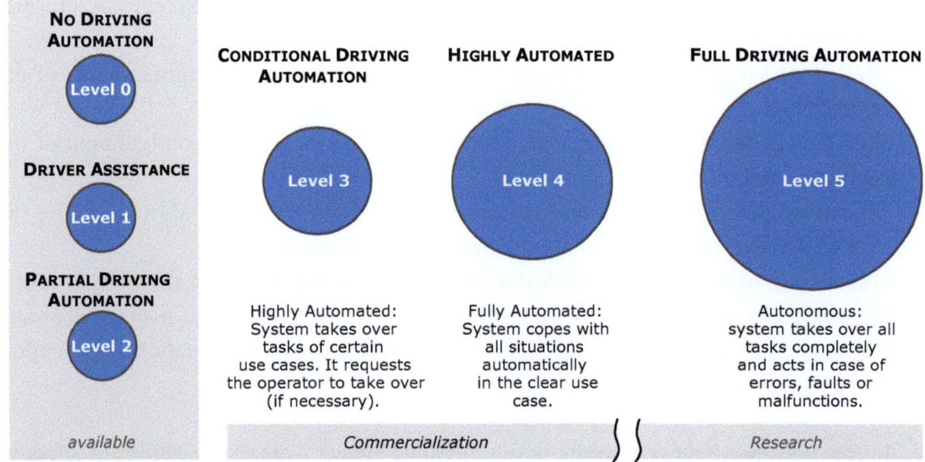

Fig. 2.13 The five levels: from manual operation to autonomous system (level definition according to SAE, 2021)

systems is at best seen in specialized applications with well-defined system capabilities. For example, fully automated mining vehicles or harvesting machines are applications with more clearly structured contexts than urban driving scenarios.

Similarly, the field involved in the automation of production systems is actively pursuing autonomous systems. Due to the shortage of skilled labor, there is a growing need to increase automation in industrial automation, for example in plants and factories, and realize autonomous capabilities. One of the goals may be to achieve "fully automated" Level 4 operations, with human intervention through teleoperation only kicking in when errors or malfunctions occur.

However, handing over control to autonomous systems represents a further step in their evolution. It is foreseeable that a societal discourse will be necessary in order to deal with systems which make autonomous decisions that could lead to unpredictable harm. While society has become accustomed to the risks associated with the operation of technical systems, many risk management issues not known to this extent beforehand will arise in the production and development of autonomous systems.

It is now becoming clear that autonomous vehicles will play a pioneering role in terms of society's acceptance of this technology. They will pave the way for discussions about legal responsibility and ethics in decision-making.

2.7   Prompts for Reflection

The chapter presented the basic concepts of automation. It gave a clear outline of the evolution of automation engineering over the past few decades from an engineering science to a control engineering interpretation of mechanical engineering systems. The concept of a technical system was initially formulated and later defined as an abstraction of the technical process. Both concepts have been established and standardized for many years now.

At present, the focus is on software-intensive systems capable of processing large amounts of data, possessing cognitive capabilities, and possibly embedding artificial intelligence. These automation systems go beyond simple programming and demonstrate the ability to operate autonomously in complex and unstructured environments.

Despite many differences, the boundaries between traditional automation engineering—involving meticulous system design by skilled professionals—and systems developed using artificial intelligence methods and capable of autonomous operation are becoming increasingly blurred.

To delve deeper into this topic, it is recommended to explore further facets, conduct additional research, and consider the potential future trajectories of automation engineering.

Here are some possible avenues you might consider:

a. Analyze how the concept of a technical process has evolved in recent years and what role the IT abstraction from a concrete technical device to a technical system or process has played in this process.
b. In this context, it is also interesting to distinguish between the terms "automation technology", "automation", and "automization" and between "automation systems", "automated systems", and "autonomous systems".
c. Consider the increasing use of software in automation. How are system structures changing and what is the influence of information and communication technology? How do you assess the use of large software systems in this context? What role will the availability of data play?
d. Form an opinion on the question as to how automation technology will evolve towards networked cognitive capabilities in the future. How do you see the future of autonomous systems and where do you see the upcoming challenges in development and operation?

Now that you have acquired some knowledge of the fundamental concepts of automation technology and its evolution, you are equipped to delve deeper into the subject. Numerous questions may arise in relation to this topic, and you now possess the necessary expertise to approach them.

Further Reading

BMWK: **Plattform Industrie 4.0 Glossary**, Berlin 2023
https://www.plattform-i40.de/IP/Navigation/EN/Industrie40/Glossary/glossary.html
Deloitte: **Data Nation Germany, The German Technology Sector. From Hardware to Software and Services**. Mai 2019

Drews, P.; Starke, G.: **Mechatronics and robotics in Europe.** Proceedings of IECON '93 - 19th Annual Conference of IEEE Industrial Electronics, 1993. https://doi.org/10.1109/IECON.1993.339115

Gamer, Th.: **Autonomous systems, autonomous collaboration.** ABB Review, 2018

Kagermann, H.; Wahlster, W.; Helbig, J. (Ed.): **Recommendations for implementing the strategic initiative INDUSTRIE 4.0.** Acatech and Forschungsunion, Final report of the Industrie 4.0 Working Group, 2013

Sifakis, J.: **Autonomous systems. An architectural characterization.** In: Boreale, M.; Corradini, F.; Loreti, M.; Pugliese, R. (eds) Models, Languages, and Tools for Concurrent and Distributed Programming. Lecture Notes in Computer Science, Vol. 11665. Springer, 2019. https://doi.org/10.1007/978-3-030-21485-2_21

References

Ashtari B. T.; Jung T.; Lindemann, B.; Sahlab, N.; Jazdi, N.; Schloegl, W. und Weyrich, M.: **An architecture of an intelligent digital twin in a cyber-physical production system.** Journal at—Automatisierungstechnik, 2019. https://doi.org/10.1515/auto-2020-0003

Carrez, F. (Hrg.): **Final architectural reference model for the IoT,** V3.0, IoT A Deliverable D1.5, EU G 257521, 2013

DIN IEC 60050-351:2014-09. **Internationales Elektrotechnisches Wörterbuch - Teil 351: Leittechnik** (IEC 60050-351), Beuth-Verlag, 2014. https://doi.org/10.31030/2159569

Holdren, J.P.; Lamder, E.; Varmus, H. (Ed.): **Designing a digital future: Federally funded research and development in networking and information technology.** Report to the President and Congress. Executive Office of the President, 2010

Lauber, R.; Göhner, P.: **Process Automation 1.** (in German) 3. Edition, Springer, 1999. https://doi.org/10.1007/978-3-642-58446-6

Lee, Ed.: **Cyber-physical systems virtual organization - Fostering collaboration among CPS professionals in academia, government, and industry.** NSF Workshop on Cyber-Physical Systems October 16–17, Austin, 2006

Müller, M.; Müller, T.; Ashtari, B.; Marks, Ph.; Jazdi, N. and Weyrich, M.: **Industrial autonomous systems: a survey on definitions, characteristics and abilities.** at - Automatisierungstechnik, vol. 69, no. 1, 2021. https://doi.org/10.1515/auto-2020-0131

SAE: **Taxonomy and definitions for terms related to driving automation systems for on-road motor vehicles.** ISO/SAE PAS 22736:2021, ISO/SAE, 2021

Sifakis, J.; Harel, D.: **Trustworthy autonomous system development.** ACM Transactions on Embedded Computing Systems, 2022. https://doi.org/10.1145/3545178

VDI: **Industrie 4.0—Terms.** Working Group „Terms" of the VDI/VDE-GMA FA 7.21. Düsseldorf, 2019

Weiser, M.: **The computer for the 21st century.** ACM SIGMOBILE Mobile Computing and Communications Review, Volume 3 Issue 3, 1999. https://doi.org/10.1145/329124.329126

How and Where Is Automation Technology Deployed? Exploring System Topologies and IT Architectures

3

Abstract

This chapter explores the application of automation technology and its widespread adoption in various industries. The discussion will concentrate on the different system topologies and IT architectures commonly used in practice. It should be noted that the use of automation technology is heavily influenced by the specific requirements of each industry, requirements which will be discussed in detail in this chapter.

Relevant examples are used to analyze and illustrate how the typical system topologies and IT architectures are used in the automation of industrial production systems, products, and cyber-physical systems.

This involves exploring questions that illuminate the contextual conditions within the application areas, thereby fostering an understanding of the structures of automation technology, such as:

- In which industries is automation technology currently being used, and what are the practical requirements?
- What are the characteristics of the system topologies used in production and product automation and in the automation of cyber-physical systems?
- Which contextual conditions within applications are responsible for practical IT architectures, and what examples can be cited?

In this context, we shall present some common explanatory models that describe the mindset of the relevant application areas of the industries in which automation technology is in use.

M. Weyrich, *Industrial Automation and Information Technology*,
https://doi.org/10.1007/978-3-662-69243-1_3

3.1   Application of Automation Technology in the Industries

Automation technology is driven by the need for efficiency and the benefits it provides. This requires collaboration between international corporations, small and medium-sized enterprises, and startups, supported by research that identifies problems and provides solutions.

However, what determines the complexity of automation technology, and how is its functionality and relevance defined? To begin with, it is notable that the demand for automation technology and the need to apply it within manufacturing sectors is also characterized by the requirements of industrial companies that produce similar products or provide similar services.

In addition, numerous industry-related services are generated within the sphere of companies that use or operate automation technology. In terms of software and hardware component vendors, machine system suppliers, and system integrators who market automation-related products, automation is a key business area.

There are two perspectives on automation technology: that of the users and that of the automation technology manufacturers who provide solutions to the users. Comprehending the use and deployment of functionality within automation systems requires an understanding of the practices and application processes within them to reveal patterns of interactions, interfaces, functional components, and system topologies.

From an information technology perspective, system topology plays a critical role in the categorization of automation systems. In addition to the basic system topology, the implementation of the information technology by means of IT architectures is of major importance. An IT architecture refers to the internal structure of individual components within an information technology system, thus delineating relationships between software components and their communication.

While the system topology sets important contextual parameters, the IT architecture defines the implementation of information technology in practical terms. Consequently, this chapter discusses various system topologies and established IT architectures in practical terms and examines their application within industries.

3.1.1   Automation Technology in the Industries

Automation systems have become a ubiquitous presence in various industries. As such, they have come to play an integral role in both our personal and professional lives. To better comprehend the functions and capabilities these systems require in different industries, it is essential to explore the fields in which they are used.

Figure 3.1 showcases several examples of industries that have embraced the use of automation technology.

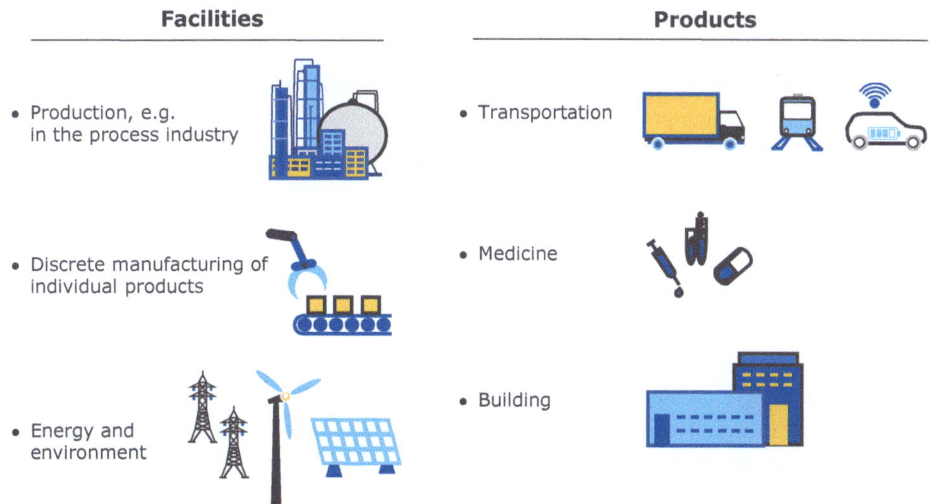

Fig. 3.1 Examples of industries using automation technology

Automation technology has a broad range of applications across various industries. In the production and manufacturing of industrial products, it is relevant in particular for high-wage industrial locations such as Germany.

Automation technology has many facets, and we distinguish between production, manufacturing processes, and application methodologies. For example, the automation of flammable chemicals—a product of the process industry—requires a different approach than the manufacturing of discrete parts such as vehicle batteries does.

Automation technology is also revolutionizing the transportation industry. There are a number of new ways in which drivers or pilots can be increasingly relieved by automation technology until they are eventually replaced by an autopilot in the near future.

Automation technology has brought about significant changes in the field of medical engineering, where it is used in data acquisition, the automated evaluation of laboratory results, and robotic support during surgery.

Buildings also utilize automation technology for heating and air conditioning, solar shading, and lighting, among other things. This provides numerous possibilities for automated functions.

Energy supply units such as power plants, wind turbines, and hydrogen systems require automation to coordinate the many distributed energy sources and energy sinks in an optimal way.

In the realm of environmental engineering, automation technology plays a critical role in protecting and restoring the environment, one example being automated information processing used to track the CO_2 emissions of industrial goods.

Depending on how, where, and for what purpose automation technology is used, users and system developers may face a variety of challenges.

There are a number of different requirements:

- Quantity and cost of the systems, which can range from complex single-purpose systems to high-volume products.
- The lifetime of the systems, which can range from a few years for consumer products to many decades in the case of capital goods. There are major differences between short-lived products and large-scale systems, in which the long-term availability of spare parts can be an important consideration.
- The qualification standards of the people using the system. This determines how the system will be operated and whether people will need to be trained beforehand.
- The power requirements, which are in particular relevant in terms of battery replacement cycles for non-stationary systems.
- The extent to which safety measures are required in relation to the risks posed by critical processes or autonomous system decisions.

Admittedly, there are some parallels and opportunities for transfer between industries, but the technical differences across industries can have significant implications for the design of technical systems, their execution, and the choice of IT architecture.

3.1.2   Classification of Automated Systems

Three categories of automated systems are used in industry, categories which can be identified according to various characteristics such as the size of the facility, the number of products manufactured, the degree of networking in terms of information technology, and the structure of the hierarchy in each case. These categories are production system automation, product automation, and cyber-physical automation.

Figure 3.2 shows sample instances of each of these categories.

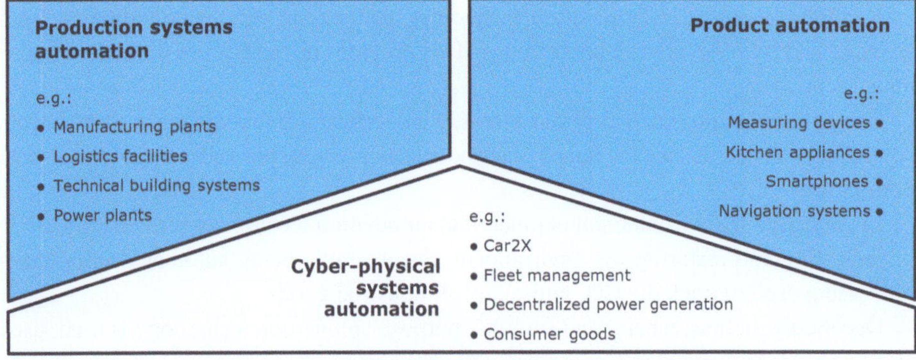

Fig. 3.2 Sample applications of the three categories of industrial automation

I. Production system automation

Production systems are composed of many devices and machines. Lauber and Göhner (1999) defined them as follows:

> The automation of production systems refers to a technical system in which the technical process is spread across different locations. These systems are characterized by a hierarchical structure involving the interaction of various components. Production systems are typically produced in small quantities or as one-of-a-kind systems.

Industrial production systems are dedicated to various domains such as manufacturing, logistics, building, power generation, and rail systems. All of the systems in this group have a large physical footprint, making an extensive and complex process and network necessary.

II. Product automation

The systems used to automate products are structured quite differently. Lauber and Göhner (1999) define them as follows:

> Product automation systems are designed to operate the technical processes in devices such as machines. These systems are well known for their compact structure and design and are used to produce large quantities of automated products.

Examples of these products can be found in various aspects of daily life and range from household and kitchen appliances to construction machines, work machines and vehicles.

While these two categories were initially sufficient to fully classify automated systems, a third category has emerged and is now becoming increasingly important.

III. Cyber-physical systems automation

The automation of cyber-physical systems is in line with the spirit of highly networked automation systems.

In essence, they can be defined as systems that integrate the cyber and physical domains:

> The automation of cyber-physical systems is based on information technology which is highly networked and seamlessly integrated into virtual ("cyber") and physical environments. Today they are fast becoming invisible and all-pervasive while developing new capabilities such as providing context-aware natural interactions and services.

These automation systems are interconnected using communication technology to constitute informationally distributed systems. In tandem with the physical manifestations of subsystems, these systems exchange extensive data that can be captured in a digital twin.

This digital representation of physical components within an information domain as articulated by Porter and Heppelmann (2014) is actualized by the integration of distributed "smart" products. Consequently, the interconnected automation system develops capabilities derived from the functions and attributes of its subsystems.

One example of this concept is automated fleet planning as used in agriculture. It involves coordinating the operation of agricultural machinery, route planning, weather forecasting, and crop yield prediction. The technical (sub)systems are controlled via interconnected data and the resulting information.

However, some automation systems may not lend themselves to immediate categorization. In these cases, the characterization depends on the nature of the overall technical system. One example that illustrates this fluid transition between the above types is the automation of food production.

A typical bread-making machine is a product automation system. It has a compact form, is mass-produced, and performs a dedicated function. In contrast, an automated bread factory in which the "baking" process is fully automated using multiple devices and transport mechanisms falls under the category of production system automation. Here, individual functions such as "measuring ingredients," "mixing dough," "forming loaves", and "baking" occur in automated machines which are interconnected and have an extension in space.

3.1.3 Requirements for Automation Systems

In industries that utilize automation systems, functional requirements are established to define the scope and the characteristics of the system. These functional requirements specify what the system should do and how it should perform. For instance, an automated heating valve should have an auto-calibration function, and a controller must implement certain specific temperature responses.

Table 3.1 summarizes the typical industrial requirements for automated systems, demonstrating that specific requirements vary depending on the industry in which the system is implemented.

Different industries have varying requirements, making it necessary to use different automation technologies. For instance, low cost and a high manufacturing volume are the focus in the production of household and consumer goods. In contrast, laboratory metrology requires development environments that are highly configurable and as universal as possible to enable rapid production of structures and flexible responses to ever-changing requirements. Vehicle systems require modularity, robustness, and convenience, while aviation demands highly reliable certified systems. Building automation, on the other hand, requires simple but durable components.

Additionally, there are diverse hardware and software requirements which are crucial from a technical, business or organizational perspective.

Table 3.1   Requirements and conditions for automation systems in the various industries

Type of industry	Functional requirements	Underlying conditions
Industrial plants, e.g. chemical industry	Integration of different certified subsystems into individual end-to-end systems	One-of-a-kind systems require specialist personnel for customization
Household appliances and commodities	"Throw-away" systems in large volumes with a demand for cost efficiency; possible replacement of modules	Repair through exchange of modules by skilled personnel
Laboratory equipment for metrology, e.g. in quality assurance	One-of-a-kind systems that are specially adapted and are to be used flexibly	Flexibility in deployment possible with expert users
Vehicles (cars and rail)	The vehicle manufacturer and selected suppliers define the modular structure	Repair through replacement of modules by specialist personnel, further development of existing systems by experts after an elaborate approval process only
Aircraft	Complex system validation due to safety-critical application	Costly certification hampers further development
Building automation	Cost pressure, modular structures and long lifecycles	Repair by qualified personnel, extensions may be required for retrofits

3.1.4   Additional Requirements for Automation Systems

Depending on the industry and the context in which a product is used, there may be additional factors to consider in addition to the initial design.

For example, in the case of a heating valve, it is important to include appropriate automation technology for control purposes in a low-cost consumer product. Also, factors such as size and power requirements will vary depending on the specific application area involved.

Figure 3.3 provides an overview of these additional requirements and related aspects.

Consider, for example, the field of technical equipment in the chemical or manufacturing industries. The maintainability of hardware and software and the rapid availability of spare parts are of great significance here. When these facilities are exported, ensuring the global availability of spare parts becomes critical for preventing downtime and the high costs associated with it. This makes it imperative to facilitate rapid on-site replacement.

Certification is of obvious importance for safety-related equipment. Control systems of this kind often operate unchanged for many years due to the complex and costly nature of recertification. In addition, the risk of failure during the implementation of new features outweighs the stability of system operation. As a result, these systems often cannot be interconnected because continuous updates would undermine certification. Periodic recertification would entail immense costs that cannot be justified by the functional gains of the overall system. In addition, from a developmental perspective, the availability of special-

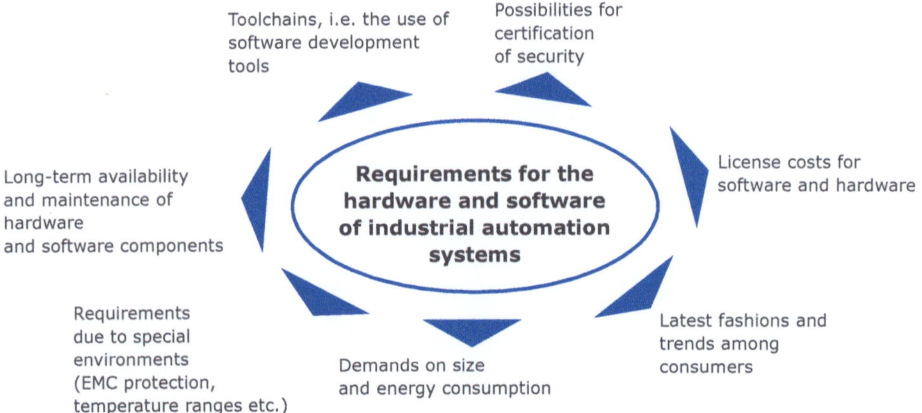

Fig. 3.3 Additional requirements for industrial automation systems

ized development environments known as "toolchains" could influence hardware and software decisions.

While it is possible to automate many processes today, an assessment must be made as to whether the development effort or the benefits of the automation technology outweigh the costs. The effective streamlining of the development process—"automating the automation", that is—opens up numerous new fields of application due to the improved cost-benefit ratio.

Conversely, automated development processes suffer a loss of flexibility. When electronic components become unavailable, or when updates and support for operating systems, compilers or other development environments are discontinued, the resulting supply bottlenecks can lead to significant constraints as entire development processes must be restarted in order to maintain costly capital assets.

Compared to the rapid development of consumer products like smartphones, the implementation of innovations in industrial automation systems is slowed down by the requirements presented here. In the case of smartphones or laptops, it is common for the support for system features to be discontinued after a few years. In the case of vehicles, machines or factory equipment, much longer operating times are required to ensure investment protection.

3.2   System Topologies and IT Architectures of Production System Automation

3.2.1   Characteristics of Production System Automation

An industrial production system is characterized by an automated end-to-end process consisting of a series of spatially distributed subprocesses and an overarching supervisory instance. A typical example of this is a chemical plant, where raw materials are processed through a number of stages to produce valuable substances.

The characteristics of the industrial automation of production systems can be described as follows:

- These systems exhibit complex processes and operations with multiple automation functions.
- Industrial production system automation is characterized by numerous sensors and actuators, controllers and regulators—typically decentralized within individual subprocesses—as well as overarching supervisory control systems.
- Manufacturing plants such as those for the discrete manufacturing of automotive components typically operate for 5 to 10 years before the production facilities are rebuilt and the equipment is reused. Chemical production plants have a much longer service life, with equipment lasting for up to 30 years.
- Industrial production systems are typically built only once, i.e. as one-of-a-kind systems, or in very small numbers, although in some cases systems are duplicated to allow production at several locations.

Given this scenario, it is evident that the overall cost of automation technology within a facility is primarily determined by the development costs. The installation of a slightly more expensive system component then makes little difference as the engineering costs outweigh expenses of this kind.

Safety is another crucial aspect of industrial automation as a production facility can pose significant hazards.

3.2.2   The Automation Pyramid

Production systems have a hierarchical structure that results from the complex interactions involved in automated production.

There are norms and standards that delineate hierarchical system topologies for various domains of application. In this context, system manufacturers and large-scale users have developed their own corporate standards that extend the general guidelines, for example in automotive or chemical production. International standards such as IEC DIN 62264, 61512, and EN 81346 shape the structure of production automation by defining the automation pyramid. Due to the historical development, industrial production systems are also referred to as ISA-95 or ISA-88 systems, which are systems based on the initial American standards.

In this context, a pyramid depicting the systems and automation technologies for production systems in the form of levels emerged in the 1980s.

The automation pyramid delineates and explains the components and interactions of industrial production.

Fig. 3.4 The automation pyramid: hierarchical topology with categorization of system levels

The automation pyramid is shown in Fig. 3.4 and is defined as follows:

The automation pyramid is a structured model that covers the production systems, from sensing and controlling the process to coordinating the process, as well as operational and corporate management.

The automation pyramid is used to describe the hierarchical topology of production system automation. We can distinguish five levels, each with a different task:

1. The field level of the pyramid is situated at the bottom, forming its base. This is where the technical process components such as sensors, actuators, and other equipment are located. It is also the place where inputs and outputs occur and process signals are provided. The field level is responsible for collecting information through sensors and applying it to the technical process through actuators. This area is physically close to or directly connected with the technical process.
2. Above the processes of the field level, there is a control level which is responsible for collecting and adapting the data. This area is used for the functional realization of the control. The components of this subsystem operate in real time and are subject to time constraints as they interact directly with the technical process.
3. At the supervisory level, information from various control processes is gathered and analyzed to monitor and manage the entire process, which may be distributed across different locations. SCADA systems at this level provide the tools necessary for collecting and analyzing data from multiple domains, primarily the control domain. This data can be processed for the purposes of transformation, storage, modeling or analysis.
4. At the manufacturing planning and operation level, work processes are organized and connected to enterprise resource planning systems. This level supports automation and analysis functions such as automated data collection, processing, and validation. It also provides data on downtimes and delays, enabling their causes to be identified. Forecasting, diagnostics, and optimization are some of the responsibilities of this level.
5. At the top of the pyramid is the enterprise level, where the initial planning of production and sales is organized. Here, raw materials and preliminary products are requested,

and their logistics must be planned for the long term. This layer represents the implementation of the business logic and consists of functions for higher-level planning.

The levels of the automation pyramid can be illustrated by the example of an automated juice bottling factory. The subsystems such as the sensors and actuators used for blending and bottling the juice are assigned to the field level. Each machine station is equipped with a PLC at the control level. A higher-level SCADA system manages the PLC at the supervisory level. At the factory level, MIS and MES are used to determine which juice to fill and when to fill it. This is where operational management occurs, and schedules are adjusted for the next few hours.

Finally, at the enterprise management level, long-term plans are made. For instance, market forecasts are used to purchase the right amount of precursor products such as juice concentrates or special bottles and arrange transportation and storage supported by ERP systems.

The automation pyramid allows for a smooth transition from concrete, machine-related functions such as processing measured values at the bottom of the pyramid to more abstract and overarching functions such as production planning at the top of it.

The automation pyramid provides a standardized way of assigning roles, tasks, and responsibilities within an industrial production system. The hierarchical structure assigns individual tasks and functions, enabling consistent classification.

3.2.3   Allocation of IT Systems to the Different Levels

In accordance with the tasks we find on different levels of the automation pyramid and the corresponding functions, practical segments for which commercial systems are available have come into being.

There are a number of typical examples of implementations. Figure 3.5 showcases the implementation of the IT architecture for the automation of industrial production systems.

1. Inputs and outputs as well as process signals are to be found at the field level. This is also where we find the sensors and actuators connected to the technical process. These components may include switches, pressure sensors, temperature sensors and so on. It is important to note that modern sensors are complex automation systems equipped with numerous controllers and sensors designed to measure and process physical variables in the process. At the field level, the individual components are connected to one another or to the control level via bus systems.
2. Programmable logic controllers (PLCs) are located on the control level of a manufacturing process. PLCs are responsible for evaluating individual signals—such as logical links like "limit switch reached, motor stopped"—or for controlling manipulated variables such as the temperature by means of a control valve.

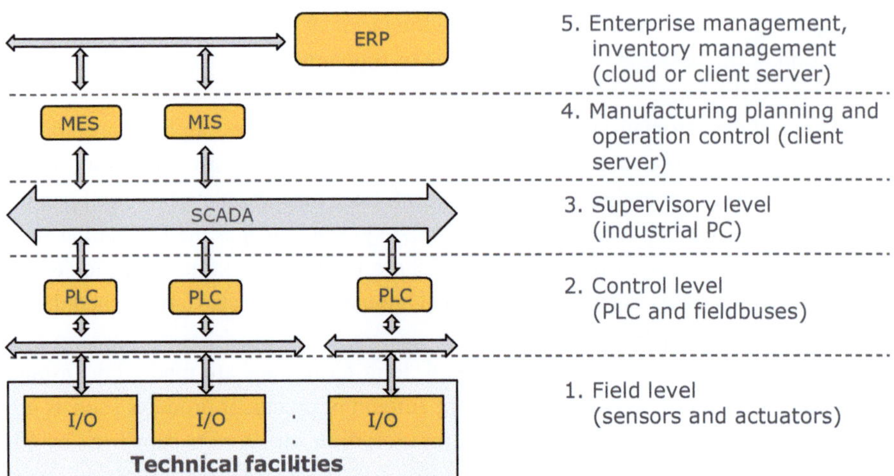

Fig. 3.5 Implementation of a system topology according to the levels of the automation pyramid

3. Operators use Supervisory Control and Data Acquisition (SCADA) systems to supervise and control processes located above the control level. SCADA systems are responsible for monitoring and handling alarms and for accessing, retrieving or evaluating data. They help us to display trends in process performance and allow human interaction by means of higher-level control and process monitoring. SCADA systems also enable functions such as predicting equipment life, supporting troubleshooting and maintenance, and providing process documentation with the help of automated reporting which meets the regulatory requirements.

4. On the operational level, two types of software systems support the coordination of operations: the Manufacturing Execution System (MES) and the Manufacturing Information System (MIS). MES systems are responsible for production planning and process optimization, including scheduling production steps and deploying personnel. Meanwhile, MIS systems collect and evaluate data from SCADA or ERP systems and present information about production to support decision-making. For instance, an MIS system can help with quality monitoring by tracking and detecting any issues early on.

5. At the enterprise management level, companies have adopted Enterprise Resource Planning (ERP) systems that allow for streamlined merchandise management and commercial processing.

The distinctions between the levels of individual systems and system environments are not always clear-cut, instead depending on the industry and the application concerned. For instance, machine tools and industrial robots have a powerful control level that manages complex motion and sequence control and an equally powerful manufacturing planning and operation level which provides appropriate tools. However, the SCADA level (process control) has little or no significance in these cases due to the limited number of monolithic units. On the other hand, the SCADA level is highly important in chemical installations or large power plants because of the distributed nature of the automation components, and it requires a lot of information to be compiled and monitored. Depending on the situation, therefore, the emphasis may be laid more on one level than on the other.

The topology of the automation pyramid system is governed by a clear hierarchy of levels. Data and information are collected in the field and then aggregated upwards in the hierarchy to form a higher-level plan. Of course, both the data structure and the time-frames used for processing it differ. In the field, data is processed in real time to allow the rapid tracking of conditions, i.e. physical processes. Process variables at the field level such as the motion control of an industrial robot occur in the microsecond or millisecond range and require direct intervention.

In contrast, tasks related to production planning and enterprise resource planning run somewhat more slowly on the upper levels, taking from several minutes to several hours.

This hierarchical topology of the automation system is constantly evolving due to the advancement of digitalization processes. Networked and distributed IT architectures in which the location of software execution is secondary are now becoming the norm. The software can be directly integrated into the facility in the form of a control unit or executed remotely in the cloud.

3.2.4   IT System Modules for Production System Automation

A wide spectrum of hardware and software products is available for industrial automation in production. Many companies offer both complete automation systems and individual IT system components as commercial products. Reliable functionality, security, and usability play a key role in this scenario.

A large number of IT system modules have been developed, and several commercial solutions are available for each of them:

– ERP software for enterprise resource planning and value chain management
– MES and MIS software for production planning and process visualization
– SCADA systems and applications with cloud platforms for diagnostics, remote mainte-
  nance, teleoperation, etc.
– Controllers (PLCs and motion control)
– Sensors and actuators, for example for drive control
– Various communication systems for connecting components and subsystems.

Depending on the segment of industry involved, a wide range of products are on offer, including specialized solutions for specific applications. Reliable functionality, safety, and ease of use are critical here.

However, the long lifespan of these systems means that technology cycles are generally slow. As a result, spare parts and systems for older facilities are still available on the market even though the underlying technology has become obsolete. Consequently, new facilities may be forced to use older generations of technology that have been on the market for years even though improved IT system components have long replaced them.

However, these new facilities could still be built with older technology to protect investments. This approach allows the implementation of uniform system components across the enterprise, simplifying spare parts management even though improved IT system components are available.

3.2.5   The Virtualization of Hierarchies

The automation pyramid depicted in Sect. 3.2.2 is inspired by the process control environment. However, these standards stem from a time when system development was driven by the implementation of automation system hardware. At that time, technical systems were characterized by an infrastructure in which the locations of the technical process in each case—such as the placement of sensors and controllers in control cabinets—significantly influenced the system topology and the IT architecture, and the physical assignment of these elements to machines was determined by cable lengths or cabinet locations.

In the early days of automation and information technology, sensor, control, and office networks were technologically distinct and not interchangeable. This led to a hierarchy in which each functional block was able to interact with adjacent blocks.

Advances in information and communication technology indicate a trend toward increased flexibility in system topology, allowing implementation concepts that transcend the spatial boundaries previously defined in the automation pyramid. Figure 3.6 outlines an IT architecture in terms of a hierarchical, distributed and decentralized topology.

This concept assumes an increasing interconnection of systems and components by means of information technology. Subsystems and their components form an IT system which is digitally integrated as an IT infrastructure like a cloud which connects the physi-

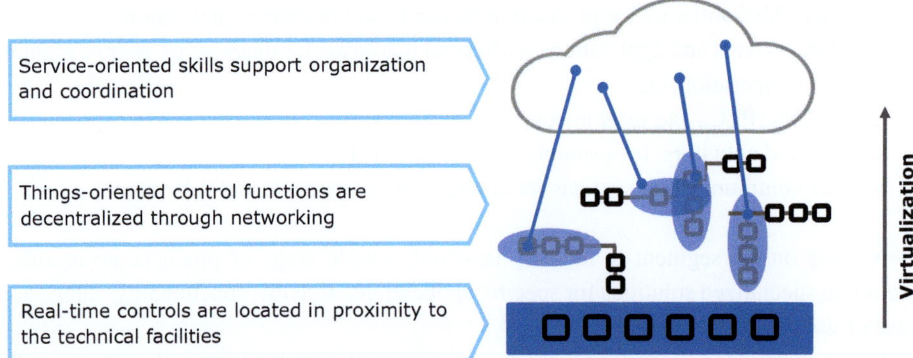

Fig. 3.6 Sketch of a virtualized IT architecture

cal components. Unlike in the past, the rationale for hierarchical structuring is no longer geographical, instead relating to the logical grouping of the function a system is intended to perform.

Virtualization involves the use of software and communication technology to realize functionality in connection with mechanical system components. It creates an IT architecture in which "services" are abstract expressions of individual software functions. These software services can then be provided by another subsystem. This scenario is referred to as virtualization since the location in which the software is executed is unimportant as long as the function in question is delivered in a timely and reliable manner. Subsystems in the technical process can be directly equipped with control technology or can access other networked subsystems to allow calculations to be performed.

As a result, technical barriers that currently require structuring into system groups based on physical proximity are eliminated. By removing these technical limitations, the analogy to the hierarchical levels of the automation pyramid becomes unnecessary, allowing previously rigid forms of organization to become more flexible technologically.

For many applications in industrial production system automation, however, a clear hierarchical organization will continue to be necessary for managing the complexity of technical facilities, but these hierarchies will be based on logical aspects instead of the physical location of an industrial installation.

Nevertheless, the operation of automated production systems is often safety-critical and is characterized by various reliability requirements, especially in terms of communication times and the IT security of communication links. In motion control systems, for example, time requirements are in the microsecond and millisecond range. Conditions of this kind often cannot be virtualized economically in industrial IT networks due to the delays occurring there. For this reason, the large-scale virtualization of time- and safety-critical control functions is not expected for the near future, and it will continue to be necessary to implement these control functions in close proximity to the technical system.

3.2.6   The Reference Architecture Model for Industrie 4.0 (RAMI 4.0)

As part of the efforts to standardize the platform known as Industrie 4.0, a new hierarchy and procedure model was developed. This resulted in the development of the Reference Architecture Model for the "Platform Industrie 4.0 (RAMI 4.0)" as defined in DIN SPEC 91345:2016-04.

The RAMI model is a helpful tool for classifying various standards and norms on the basis of different levels, making it easier to locate standardization activities. It essentially serves as a meta-modeling tool.

Figure 3.7 illustrates the RAMI 4.0 model, which includes three dimensions: a "Lifecycle & value stream" dimension, a "Hierarchy" dimension that aligns with the levels of the automation pyramid (hierarchy levels), and a third dimension called "Layers".

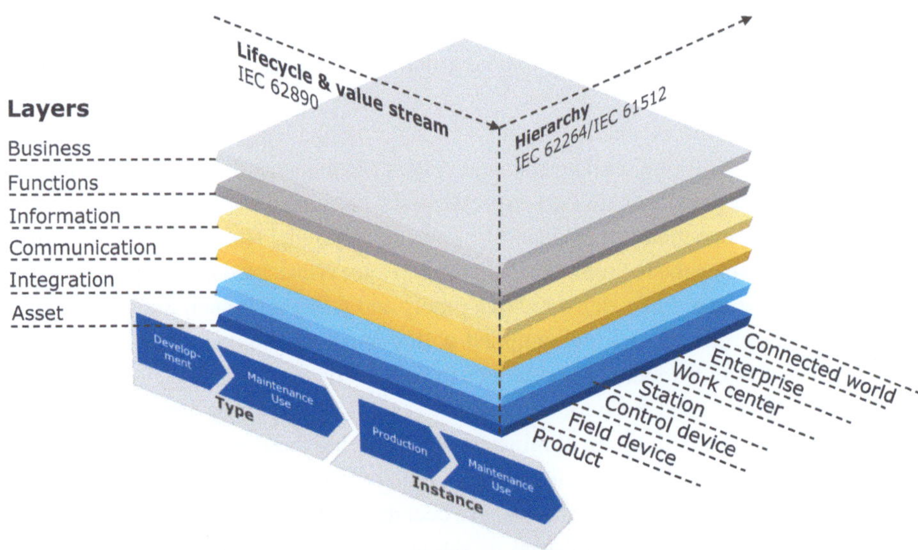

Fig. 3.7 Industrie 4.0 (RAMI 4.0) according to DIN SPEC 91345 (2016) with permission of the Plattform Industrie 4.0 and ZVEI, 2024. All Rights Reserved

Together, these dimensions create a layered cube which is also known as the RAMI 4.0 cube.

The RAMI 4.0 model incorporates relevant international standards for the automation of industrial production systems, enhancing some and placing others in a new context.

– The "Lifecycle & value stream" dimension incorporates the international standard IEC 62890 "Lifecycle Management of Systems and Components". It considers the different phases of the lifecycle, which include development, acceptance, operation, maintenance, and scrapping. In this way, it becomes possible to make a distinction between the communication of a field device in the "development" phase and its communication in the "operation" phase. This dimension is crucial for distinguishing between product creation, commissioning, the various generations of use in operation, and decommissioning. It provides additional information that complements the other two axes.
– The "Hierarchy" dimension refers to the automation pyramid, which we discussed above, but in a modified version. This dimension goes beyond the DIN EN 62264 or IEC 62264 series of standards and also includes the levels "Products" and "Connected World". Also, the DIN EN 61512 or IEC 61512 series of standards defines batch-oriented operation—also known as batch control—which is typically used in process engineering. This standard outlines process and control models and defines procedures that are also relevant in the context of the RAMI 4.0 model environment.
– There are six horizontal layers ranging from "Assets" to "Business processes" which were not defined in any of the previous standards. The "Assets" layer represents physi-

cal elements whereas the "Integration" layer provides a platform for connecting individual objects to the digital world. The "Communication" layer ensures the exchange of data and continues the connection. The "Information" layer collects all process-relevant information in various formats in which the data of the components can be stored. The "Function" layer defines tasks and responsibilities in an abstract way, for example by means of work plans or flowcharts. Finally, the "Business process" layer models business goals.

This highly advanced RAMI 4.0 model integrates and correlates the three different perspectives of automation technology—hierarchy levels, lifecycle, and value creation. Each axis of the RAMI 4.0 cube can be used independently, just like it could in the automation pyramid concept, but the axes can also be combined to allow multidimensional mapping.

Consider the example of a heating valve and its placement within the RAMI Cube. It can be classified as either a "field device" or a "product" on the "Hierarchy" axis. It may also have integrated parts of the "Control" aspect. The question now is whether the heating valve should be clearly assigned to a single level or whether components could be assigned to other levels too.

The second axis concerns "Lifecycle & value stream". Here the question arises as to which phase in the lifecycle should be considered. Is it the development environments for heating valves or their commissioning in the field? Is it remote maintenance during operation or repair processes? This axis allows assessments to be made across the entire lifecycle as well as assessments which focus on specific phases.

Finally, the assessment can be layered. Initially, the heating valve can be considered as an "Asset". However, how should the electronics be designed if they are to support integration into an IT architecture? How is "Communication" implemented, and in which way is "Information" related to the setpoints and operating statuses communicated? Which specific "Functions" should the heating valve perform? How should additional functions such as condition monitoring or safety mechanisms be handled? Finally, at the "Business Process" level, questions about revenue generation and the business model arise. Several concepts are possible, ranging from the traditional sale of hardware components and the provision of condition monitoring to pay-per-use models that include maintenance and upkeep.

This example shows how complex the structure of automation systems has become and the importance of looking at them from different perspectives. The RAMI 4.0 Cube provides a qualified and standardized view of automated technical systems within their lifecycle. By bringing the different dimensions together in a layered cube, we start to see new relationships that should be viewed together rather than separately.

3.3 IT Architectures for the Automation of Products

Product automation refers to systems in which a technical process is automated and contained within a specific unit such as a machine or device. Systems used in product automation are diverse and span a wide spectrum, ranging from household appliances like washing machines, coffee makers, and bread machines to items of everyday use like wristwatches, and extending to transport vehicles such as automobiles.

Several factors distinguish product automation, such as the level of complexity, the degree of customization, and the amount of human interaction required.

- Automation is usually performed using microcontrollers or dedicated electronics "deeply embedded" into a product in some way.
- In product automation systems, device control is implemented by means of a small number of sensors and actuators.
- Product automation is characterized by dedicated automation functions with defined system limits so that a high degree of automation is achieved.
- Product automation systems are typically produced in large volumes and manufactured in series or even mass-produced.

When it comes to manufacturing, the costs of production are weightier than the development expenses. This is because development costs can be distributed over multiple units, keeping unit costs relatively low. For example, opting for a microcontroller that is cheaper to produce per unit but requires more time for assembly language programming to ensure optimal performance may result in higher development costs. However, careful planning of high-volume production can offset these costs and lead to cost savings.

3.3.1 IT Architecture of Systems for the Automation of Products

The diagram in Fig. 3.8 illustrates how product automation systems are structured. In a basic setup, all system components are connected to a single control unit. This unit serves as the primary hub, housing both a microprocessor and microcontrollers responsible for input and output. Pre-control algorithms run on the microcontroller, which contains specialized logic circuits designed for control purposes. These units are commonly referred to as Electronic Control Units (ECUs). Nowadays, a combination of microcontroller and software is used for programming and configuration.

The controller receives signals from sensors and sends manipulated variables to actuators. Displays such as screens communicate information to the user. Input systems like touch screens or keys collect data from users. Users provide instructions, establish setpoints, trigger automated functions, and receive system status information.

The architecture of a product automation system can be intricate, often comprising several subsystems that function independently while exchanging data amongst themselves. These subsystems are connected through a network as illustrated in Fig. 3.9.

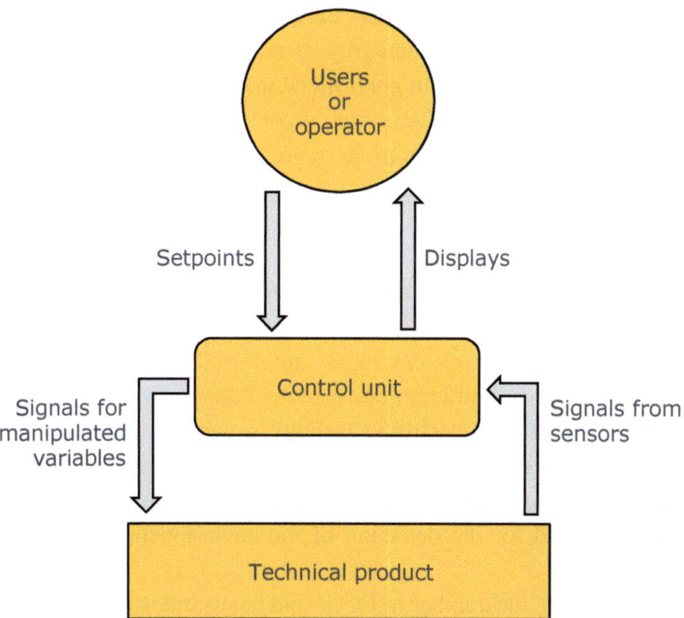

Fig. 3.8 IT architecture of a simple product automation system (based on Lauber and Göhner, 1999; courtesy of © Springer-Verlag GmbH Germany 2024. All Rights Reserved)

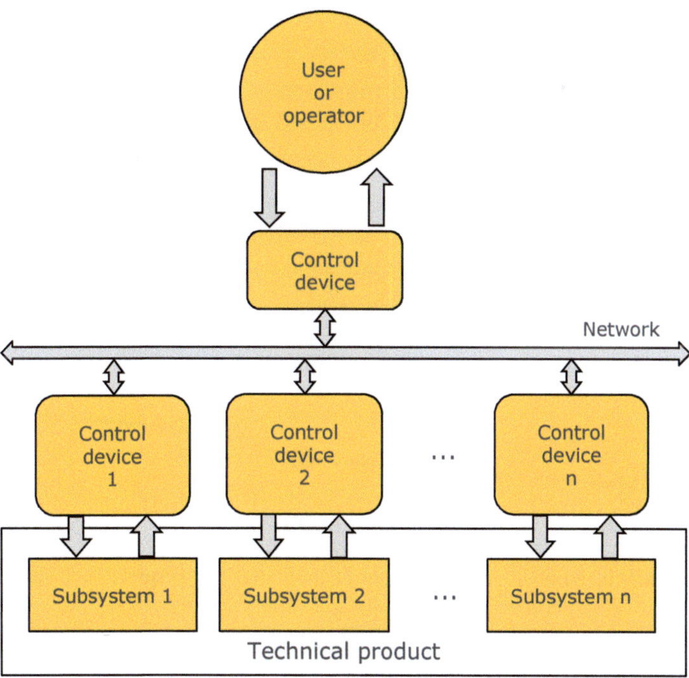

Fig. 3.9 IT architecture of large-scale product automation systems (based on Lauber and Göhner, 1999; courtesy of © Springer-Verlag GmbH Germany 2024. All Rights Reserved)

For various reasons, products are often subdivided into subsystems. Typically, these subsystems follow the functional modularity of the product and allow individual functional units to form subsystems with controllers and related sensors and actuators. The structure is usually designed to enable a central control unit to interact with the user and coordinate other control units. However, there are numerous other ways to organize the interaction between controllers, sensors, actuators, and the user.

3.3.2 Examples for the Automation of Products

The automobile is a prime example of a highly automated product. Modern vehicles contain several control units with different bus systems. Many vehicle system functions such as anti-lock braking systems, electronic stability programs, lighting systems, and an increasing number of driver assistance systems, automate several functions in the vehicle. Due to the development towards semi-automated and autonomous driving, the number of signals and sensors used for the detection of the environment is expected to increase significantly.

As shown in Fig. 3.10, the number of ECUs and bus systems in a vehicle has increased over the last few years.

In recent decades, the use of electronic control units (ECUs) in vehicles has seen a steady rise. Historically, this was often tied to the development of specific automotive

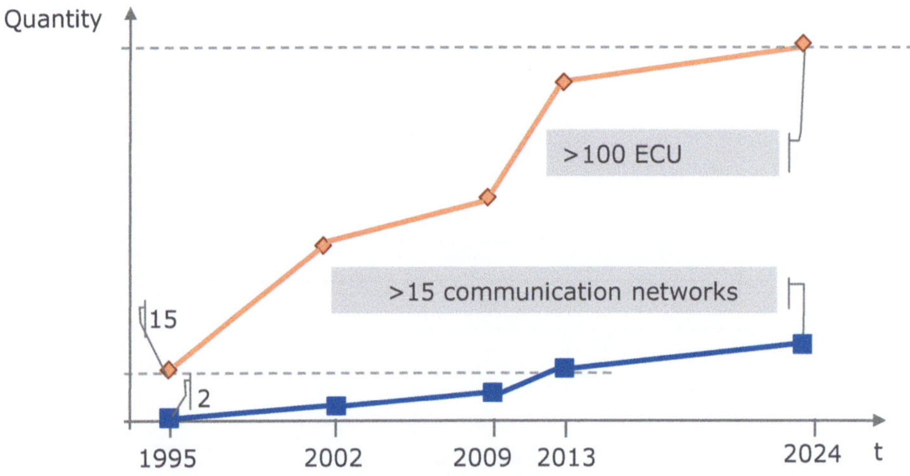

Fig. 3.10 Growth in number of vehicle control units and bus systems as a measure of system complexity

functions. However, space constraints in the vehicle, cost considerations, and the proliferation of signals have led to the consolidation of control functions and the optimization of the number of control units as well as the introduction of powerful, high-performance computing systems.

At the same time, the number of communication networks has increased significantly in recent years. This expansion has been driven by the need for safety-critical communication systems while simultaneously requiring simple and cost-effective systems. These conflicting requirements cannot be unified in the current state of the art and require several different systems to be used.

Systems with numerous components pose a number of special challenges for development. The integration of many different subsystems is difficult to coordinate. It is expected that multiple development teams across organizational boundaries will be involved in system development. For this reason, guidelines for the design of individual components and their IT integration are crucial.

Consequently, collaborative development partnerships have become increasingly important in recent years for modularizing products and establishing related industry standards and development conventions. The participating industry organizations are actively contributing to these efforts.

3.4   IT Architectures in Cyber-Physical System Automation

A continuous communication link between operational technical systems and a back end establishes information exchange and enables new coordination possibilities in a virtualized data space. The automation of cyber-physical systems becomes increasingly feasible through networked communication and data spaces. This automation of cyber-physical systems can also refer to systems of systems, where capabilities and resources are collectively managed and deployed in the physical world and in a data space. When such systems are spatially distributed and use the Internet as their enabling technology, they are referred to as Internet of Things (IoT) systems.

The potential of distributed and highly networked cyber-physical systems is extensive and includes the coordination of vehicle and aircraft fleets, applications in energy generation or distribution, the management of machine fleets, and much more.

In agriculture, significant improvements in yield can be achieved by coordinating farm machinery and planting, growing, fertilizing, and harvesting in a holistic way. By better aligning the use of harvesting machinery and processing systems with weather forecasts, crops can be harvested more quickly and with higher yields. This overarching coordination allows the production and processing of agricultural commodities to be managed more economically and efficiently.

There are three types of networking and coordination:

- Type 1: Simply connected systems: Connected products can be queried using a cloud and an application.
- Type 2: Coordination among connected systems: Individual systems share information, resulting in basic coordination.
- Type 3: Systems of systems connecting different system domains: Coordination occurs over a network with many participants and is complex and automated.

The first stage, in which individual objects communicate with a central control center, is already a reality in many fields of life and manufacturing. For example, package tracking is now a standard technological function in logistics. Similarly, monitoring the operating status of manufacturing machines and equipment in the field has become a practical application. This approach enables analysis that allows the immediate intervention into, or even the prediction of, future system conditions.

In the second stage of coordinated systems, the information obtained can be used to assess the availability of machines such as harvesters and transport equipment, thus facilitating overall optimization.

In the third stage (systems of systems), a wide range of individual product systems and additional information can be coordinated. For harvesters, this means using information on the availability of machine fleets, integrating weather forecasts, selecting seed types, and implementing systematic irrigation and fertilization. The integration of all of these individual systems allows an optimal overall result to be achieved.

The following aspects outline the characteristic criteria for networking:

- The spatial arrangement of objects or subsystems
- The quantity of systems that function concurrently
- The information technology topology governing the linkage between subsystems

The way in which these components will interact with each other is often not fully predictable during the development phase. The resulting complexity as well as unforeseen combinations of subsystems create new challenges for system development.

3.4.1 IT Architectures for the Automation of Cyber-Physical Systems

Ensuring the seamless virtual integration of objects, components, and systems requires IT architectures which are sufficiently open and flexible, even when all of the subelements come from different vendors.

It is expected that connections will be established between a variable number of systems, thus forming ad hoc networks for information exchange. This requires "elasticity", i.e. the operating platform must be able to adapt to changing demands on computing and storage resources.

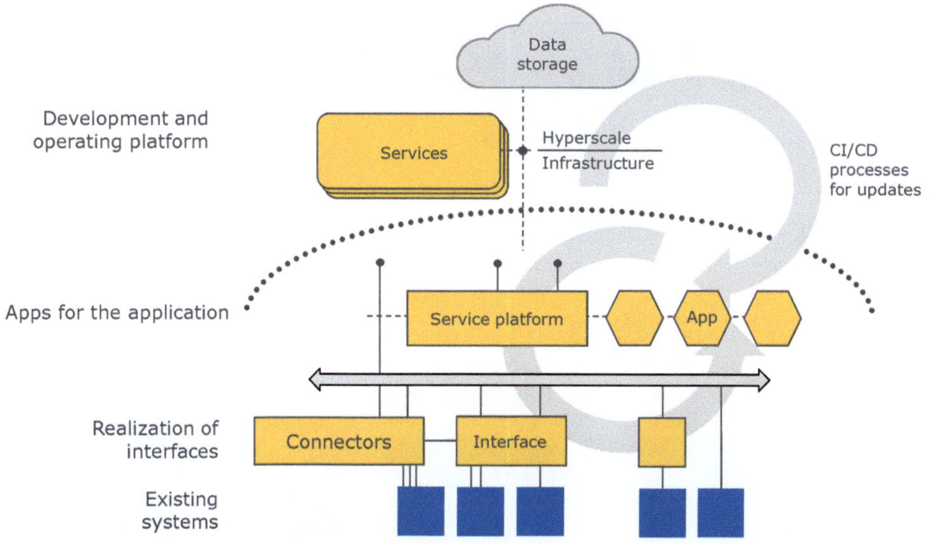

Fig. 3.11 Schematic representation of the IT architecture of cyber-physical systems in automation

It is also obvious that systems will evolve throughout their operational lifetime. Adequate performance reserves should be provided to accommodate potential expansions.

Figure 3.11 shows a schematic IT architecture of an automated cyber-physical system. In it, we can see different levels of abstraction with different implementations. Existing systems. i.e. their hardware and software implementations, are interconnected by so-called connectors. These connectors are essentially software components that organize interfaces and enable connectivity. Users can then use the functions of the cyber-physical systems automation at the next level via applications running on a service platform. These are services and data storage provided by an operating platform, typically a large IT provider (often known as a hyperscaler).

At the same time, the operating platform manages the development and enhancement of the existing software. This enables the continuous integration of new software releases, which can be implemented in the facility via a data link to the automation system.

3.4.2   Example: Car-to-Car Communication

One example of the automation of cyber-physical systems with variable participants is the communication between modern vehicles (known as car-to-car or car-to-infrastructure communication). Vehicles issue warnings or make automated maneuvering decisions based on networked information received from other vehicles via an IT network and stored in a data space.

Fig. 3.12 Connected driving as an example of cyber-physical system automation

Figure 3.12 illustrates how information about traffic incidents is exchanged between driver assistance systems. In this scenario, a vehicle's electronic stability program detects an oil spill at a highway exit. The goal is to warn vehicles and people about to use the same highway exit. To accomplish this, the vehicle communicates information about the hazard zone to potentially affected vehicles further behind.

However, this is not a simple task. As shown in the figure, many vehicles close to the hazard zone are not affected because they won't be using the exit. Warning vehicles in oncoming traffic or vehicles passing nearby would result in numerous false alarms, thus significantly undermining confidence in the warning function. For targeted communication, it is critical to take not only the geographical location of the vehicles, but also their navigation plans into account.

The structure of cyber-physical automation systems involves the individual processing of sensor data in each controller and the targeted exchange of data with the participants. This results in an IT architecture which requires controlled communication with selected participants after an event. Only when traffic is disrupted does the vehicle in question inform others that may be affected.

How can the necessary functionalities be implemented across many vehicles in the area? Several practical issues arise here. Cyclical polling of all vehicles with continuous evaluation of all driving data based on fixed rules is hardly feasible due to the real-time requirement. In addition, ensuring secure device management, identity, and access control is critical for maintaining aspects of information security. The issue of communication between segments is critical too. It is impractical to have a central control processing all the information throughout Germany or even Europe. Instead, this processing must take place within the respective local segments, which are constantly being traversed by other vehicles in a dynamic and relatively unpredictable way. Purely local radio communication between vehicles is also challenging due to range limitations.

Privacy considerations are essential as vehicle data such as those used in navigation planning may reveal confidential information about drivers and their environment.

While these system functionalities may seem straightforward and prototype-ready for a small number of vehicles, this example highlights the challenges of scalability, especially when many vehicles of different brands are involved across Europe.

3.4.3   Example: Value Networks Based on Cyber-Physical System Automation

To establish supply chain traceability and determine the carbon footprint of industrial products, managing data in the value chain of industrial production is crucial. This requires a scalable data space that connects all partners within the value network and documents information about the respective upstream products to make them traceable.

Figure 3.13 provides an example of the partners involved in an industrial manufacturing supply chain. The term "ecosystem" is used to describe value systems of this kind and highlight the interdependencies of the various companies.

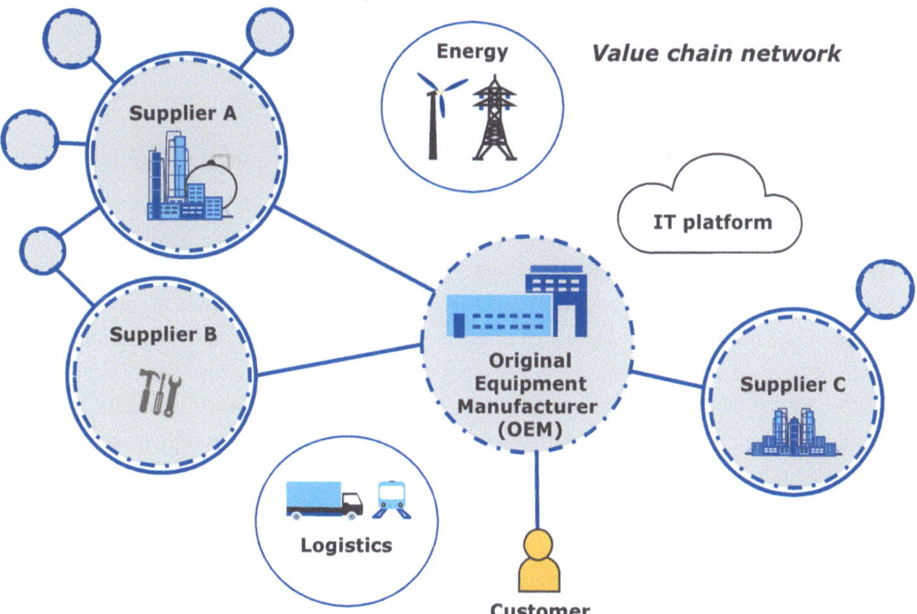

Fig. 3.13 Partner of an automatic value chain network

In this scenario, there are different products from different suppliers, comprehensive logistics, an energy supply, different production sites, IT services etc. What is needed is an IT environment that enables the creation, operation, collaboration, and automated use of all data and information throughout the system.

Nowadays, supply chains are often tightly organized and have minimal monitoring. This is due to the lack of technical and organizational capabilities for the communication-based networking of the numerous subsystems and components. There are organizational barriers to comprehensive data collection and analysis as storing sensitive information in a central database system would interfere with multiple business interests. In particular, individual partners often use different IT systems, and interconnecting these requires considerable effort.

To transition from a rigid supply chain to an agile and transparent value network based on a shared data space, both technical and organizational solutions are necessary, including:

- Connected system and data spaces with interoperable services and integration of heterogeneous and mobile hardware and software technologies
- A secure and trusted data infrastructure for information integration ensuring real-time control between domains
- The enablement of knowledge-based management decisions based on networked intelligence to protect the economic interests of participants
- The clarification of organizational aspects governing the sharing of information and responsibilities among participants.

To effectively manage the data space of the value chain, it is important to exchange information across company boundaries. During this process, the interests of all partners should be considered and accepted as being economically effective.

Here are two examples that are already partially a reality:

- To determine the carbon footprint of a technical product, it is necessary to have comprehensive information about the conditions under which it was produced and transported throughout the production chain. For the steel used in the product, for instance, details of the conditions under which the ore was mined, transported, and processed must be specified. A CO_2 label can only be reliable if all relevant information is available through track-and-trace in a data space, thus ensuring that no concealment takes place. However, suppliers may have legitimate business interests that could be undermined by the complete disclosure of production details even though transparency of this kind would be appropriate in some cases.
- In the future, truck fleets will have the ability to autonomously select a route on the basis of traffic and demand. This is made possible through networked intelligence, in which many trucks are interconnected to allow the sharing of information about traffic conditions and the incorporation of future forecasts into route planning. For instance, real-time traffic updates and historical data can be leveraged to identify congestion or

maintenance issues and suggest alternative routes and schedules. Additionally, information can be exchanged between trucks and manufacturing sites to automatically notify them of any delays and allow for rescheduling.

It is evident that this form of massive information processing in the data space holds significant potential for new system functionalities.

In terms of technical implementation, however, the question is how uniform standards for data processing can be established. Considerable effort is required in order to create an infrastructure of this kind consisting of an IT platform and the integration of participants through standardized interfaces. The development of algorithms for analyzing this networked information also raises many questions. While the benefits are often conceptually clear, quantifying them in advance is challenging.

The added value of a data space is therefore crucial. There are already numerous examples of interesting functionalities that have not been put into practice due to the lack of a compelling value proposition and the excessive effort involved. However, the threshold for the large-scale use and integration of data is steadily decreasing, and it is expected that a wide range of applications will become feasible, even if the technical effort may be considerable. In this regard, there are also many societal issues related to the collection and processing of data.

The risks associated with large-scale data collection and exchange are now beginning to emerge. It is becoming clear that, given the different cultural norms and standards prevailing across continents, different regulations and policies are likely to emerge.

3.5   Prompts for Reflection

This chapter has outlined the different approaches within the industries that use automation technology. It has become clear that, despite their having many things in common, standards and norms differ significantly. Nevertheless, typical system topologies of technical systems can be divided into three classifications: the automation of production systems, product automation, and the automation of cyber-physical systems.

This chapter provides a basic introduction to each of these system topologies and their typical IT architectures. Equipped with this information as well as explanatory models, readers will be in a position to discuss the application of automation technology across industries. This will include an understanding of the potential applications of IT architectures.

After reading this, you will be able to classify automation systems and consider categorizations and development approaches on the basis of the basic structures presented.

The following questions can help you to deepen the knowledge gained from this chapter:

Consider the reasons why you should classify the following systems as production system automation, product automation, or cyber-physical system automation.

a. A chemical reactor, an electronic stability program (ESP), an anti-lock braking system (ABS), a smart home system, a digital camera, a washing machine, a baggage conveyor system, a high-bay warehouse, a traffic light system, a parking guidance system, an intelligent sensor, and fleet management in logistics.
b. Which system topology and which corresponding IT architecture do you recommend for each implementation?

How do new system topologies and IT architectures provide the flexibility required to automate industrial processes?

Consider a modular component for flow control in a chemical plant. Given is a module consisting of a valve addressed by a servomotor to control flow. A higher-level drive controller uses interfaces to measure the flow and then adjust the setpoints.

c. Determine where the above system fits into the hierarchy.
d. How can the dimensional "layers" of the RAMI 4.0 model be assigned, and what are the transitions to the higher level or the lower level?
e. How do the topology of the technical system and the IT architecture for automation influence each other?

Consider how to build IT architectures for product automation using existing cloud platforms.

f. What are the appropriate architectures? To what extent are existing cloud platforms suitable for or adaptable to the specific circumstances of the sample deployment?
g. What is your assessment of the transition between the categories of product automation, production system automation, and cyber-physical system automation?

Conceptualize the system topology and IT architecture required for tracking high-value products.

Consider how products such as expensive pharmaceuticals can be tracked in global commodity flows. Then concretize the following aspects using a product scenario from the field of high-value pharmaceuticals:

h. What system topology is suitable for the logistics application of monitoring the distribution process and ensuring that each product can be uniquely identified and tracked in the future?
i. What are the basic IT functions and architectures required to capture and evaluate knowledge about the condition and location of the product?

Further Reading

Dotoli, M.; Fay, A.; Miśkowicz, M.; Seatzu, C.: **An overview of current technologies and emerging trends in factory automation.** International Journal of Production Research, 2019. https://doi.org/10.1080/00207543.2018.1510558

El Maraghy, H.: **Flexible and reconfigurable manufacturing systems paradigms.** International Journal of Flexible Manufacturing Systems, Vol. 17, 2006. https://doi.org/10.1007/s10696-006-9028-7

Leitão, P.; Colombo, A.; Karnouskos, S.: **Industrial automation based on cyber-physical systems technologies: Prototype implementations and challenges.** Computers in Industry, Volume 81, Elsevier, Amsterdam, 2016 https://doi.org/10.1016/j.compind.2015.08.004

Murphy, B.; Durand, J.; Lin S.-W.; Martin, R.; Mellor S. J.: **Industrial Internet reference Architecture (IIRA). Industrial Internet Consortium.** 2015

Vogel-Heuser, B.; Bauernhansl, Th.; ten Hompel, M. (Hrsg.): **Handbook Industrie 4.0, Volume 4** (in German). Springer Verlag 2017 https://doi.org/10.1007/978-3-662-53254-6

Vogel-Heuser, B.; Diedrich, Ch.; Fay, A.; Jeschke, J.; Kowalewski, S.; Wollschläger, M.: **Challenges for software engineering in automation.** Journal of Software Engineering and Applications. SCIRP, Vol. 7, Nr. 5, 2014 https://doi.org/10.4236/jsea.2014.75041

Weyrich, M.; Ebert, C.: **Reference architectures for the Internet of Things.** IEEE Software, Vol. 33-1, 2015 https://doi.org/10.1109/MS.2016.20

References

IEC 61512-1: Batch control –Part 1: Models and terminology 1997, German version EN 61512-1: 1999 Beuth Verlag, 2000. https://doi.org/10.31030/8529860

IEC 62264-5: Enterprise-control system integration - Part 5: Business to manufacturing transactions (); English version of DIN EN 62264, 2016 Beuth Verlag, 2017. https://doi.org/10.31030/2630991

DIN SPEC 91345 Reference Architecture Model Industrie (RAMI **4.0**), Beuth-Verlag, 2016 https://doi.org/10.31030/2436156

Lauber, R.; Göhner, P.: **Process Automation 1.** (in German) 3. Edition, Springer, 1999. https://doi.org/10.1007/978-3-642-58446-6

Porter, M. E.; Heppelmann, J. E.: **How smart, connected products are transforming competition.** Harvard Business Review, 2014

Software for Industrial Automation Systems

4

Abstract

This chapter provides an overview of the evolution of software for industrial automation systems.

It starts by exploring the process of creating software for automation and the important role that real-time software plays in this context. It continues with the implementation of software for control purposes in production system automation, product automation, and cyber-physical system automation, explaining the methods and procedures used. In this area, solution concepts for software development based on different approaches are illustrated in selected application areas, culminating in an examination of fundamental architectural paradigms.

Readers who engage with this chapter will gain insight into methods, procedures, and solutions to the following questions:

- How are automation systems implemented using software in industrial environments, and what are the challenges associated with real-time software?
- What programming languages and concepts are used in industrial practice to implement systems in industrial facilities, products, and cyber-physical automation?
- What basic control paradigms are available and how are software-based automation system architectures implemented with their help?

This chapter explains how modern software development goes beyond the mere programming of sequences and outlines the development approaches prevalent in this field.

M. Weyrich, *Industrial Automation and Information Technology*,
https://doi.org/10.1007/978-3-662-69243-1_4

4.1　Software for Automation

Software is increasingly becoming an integral element in the design of automation systems, leading to a significant shift in development methodologies due to the major differences between the design of mechanical systems and that of software-based systems. Unlike the visible progress observed in the construction of a machine or industrial facility, progress in software development remains largely imperceptible and is inherently more difficult to evaluate.

As a result, there has been a paradigm shift within systems engineering because of the intangible nature of software development. Software requires considerable effort for its design, implementation, and subsequent maintenance during operation. It requires thorough requirements analysis, model management, and the maintenance of source code across multiple configurations for maintenance and testing.

As a result, questions arise regarding the industrial creation of software for automation technology and the established methods and procedures.

4.1.1　How Are We to Program Automation Systems Today?

Software development encompasses far more than just the programming of a single control program that for example coordinates the sequence of operations of a machine in the form of a set of commands.

As a result, the methods and tools used to create software are extremely diverse. They range from the programming of inexpensive and simple hardware to building repetitive constellations and even large software systems. These approaches include machine-level coding, where the logic gates of a chip are linked by software, software development in high-level languages, graphically interactive program generators, and simulation-based commissioning.

Development platforms offer comprehensive capabilities which go beyond just being a programming environment, facilitating software design, system configuration orchestration, and testing.

One question remains, however: Where and why is each method deployed in industrial software development?

The approaches used to create automation systems are summarized in Table 4.1.

Machine-level coding is prevalent in areas where processing time is critical or the microcontrollers selected are simple and have some limitations. With machine-level coding, logic gates can be linked at the chip level to efficiently encode sequences, thereby automating products cost-effectively.

Software development can also be done in higher-level programming languages, languages which are particularly common in product automation.

Table 4.1 Approaches for the creation of software for automation

Approach	Procedures	Comments
Machine level and high-efficiency programming	Manual creation of Assembler and machine codes	Hardware-related programming of logic
Control software creation	Use of high-level programming languages and real-time operating systems	Design of complex control systems based on microcontrollers
Automatic code generation with models	Graphic interactive generation of software	With the help of ready-made system elements and models, the software is configured graphically and designed interactively.
Virtual commissioning	Simulation of the entire automated system and its overall behavior	The software is tested in simulation using models

For industrial process control applications, there are many graphical programming systems that create interactive models and automatically generate program code. Developments of this kind use code generators, which are increasingly being put to use in high-volume products.

In addition, vendors of software development environments typically provide complementary engineering system simulation and testing tools that allow the automated system to be tested against the generated automation software in simulation. Basic virtual commissioning tools, for example, abstract the behavior of the technical process within a logic-based simulation, replicating system behavior based on inputs and outputs and their logical associations.

In addition to 3D graphics and a physical model, complex simulators can replicate dynamic system behavior, for example by simulating a spring-damper system of a vehicle's suspension, maneuvering it around a virtual test track, and conducting trials.

These examples exemplify the diversity of approaches to software development.

4.1.2   What Is Special about Real-Time Software?

In automation systems used for technical process control, software must react in real time. According to DIN IEC 60050-351 (2013), real time can be understood as follows:

> The capability of a (process) computer system to keep the tasks in a runnable state so that they are able to react to technical process events within a predetermined time interval.

Developing software that automates in real time requires programs to conform to the same time constraints. Regardless of whether the algorithms used are simple or computationally intensive, operations must be executed neither too early nor too late. In other words, there can be no time expansion or contraction. A timeline violation of an automation system can result in malfunctions. For instance, if a sensor fails to transmit current values in real time, the automation system will be unable to perform its designated function accurately.

The term "timeliness" is commonly understood to means temporal correctness:

Timeliness is when events are responded to within a defined time period or when events are triggered at a defined point in time, i.e., within a certain time interval.

This leads to the important definition of latency (also known as response or waiting time) in automation systems (according to ISO 2382-12):

Latency is the time interval between the instant at which a controller initiates a call for data and the time instant at which the actual transfer of the data starts

Figure 4.1 depicts the time interval—constrained by upper and lower bounds—within which the latency time (T_L) of a controller must act to ensure that the automation system responds in a timely manner to events occurring within the technical process.

The consequences of not adhering to time constraints can have varying degrees of impact ranging from a decline in service quality to the complete failure of the automation system.

In a "soft real-time" system, longer response times are acceptable to a certain extent. For example, when a destination address is entered into a navigation system, a longer search time may be tolerated. Even though this may result in unwanted latency, exceeding the time limit does not pose a critical threat to the system. In soft real-time systems, latency is seen as a characteristic of poor quality, but it does not lead to fatal consequences.

In a system with "hard" real-time requirements, immediate activity is expected in the form of an activity threshold. For instance, if a traffic light turns red too late, accidents can be the result. Systems of this kind in which a missed deadline results in system failure are referred to as "hard real-time" systems.

When multiple operations occur simultaneously within these critical real-time processes, finding a way to consistently handle the processes in a timely manner becomes

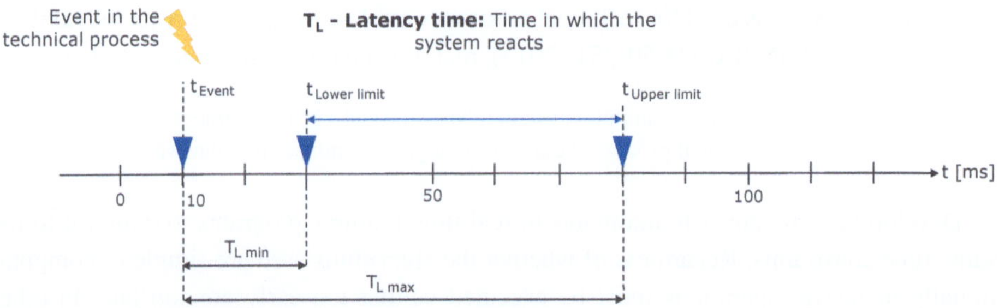

Fig. 4.1 Latency bounds for timely process operation

essential for meeting the demands. For instance, in a traffic light system at an intersection, multiple traffic lights have to be synchronized simultaneously. While it may seem that multiple operations should be handled simultaneously and concurrently by real-time software, in practice it is more practical to perform these operations rapidly one after the other. This results in the appearance of simultaneous execution although the operations are in fact performed sequentially one after the other.

Quasi-simultaneity arises in a sequential program flow when its timing adheres to the condition:

$$T_{\text{environment}} >> T_{\text{programs running in the computer}}$$

This explains why, even with sequential execution of the control program, the illusion of simultaneity is created when the traffic lights are controlled in the range of microseconds or several milliseconds, i.e. when they evoke the impression of concurrency even when executed sequentially.

4.2   Software for Production System Automation

The development of software for industrial production system automation has gained considerable importance over the past few decades. However, different domains in the industry use vastly different methods and tools to meet the diverse requirements for realizing these functions.

4.2.1   Programmable Logic Controllers

Programmable Logic Controllers (PLCs) are widely used in industrial automation. These controllers can be programmed using development environments, are used to control various systems in industrial production systems automation, and are often installed in ruggedized control cabinets.

According to EN 61131, Part 1, a PLC is defined as follows:

A PLC is a digitally operating electronic system used in industrial environments to implement specific functions (such as logic control, timing, counting, and arithmetical functions) to control various types of machines and processes by means of digital or analog input and output signals.

The various methods of programming PLCs allow the operator to quickly become familiar with programming. Both graphical and textual descriptions are used in project planning and programming. The IEC 61131, Part 3, and DIN ISO 60848 standards define textual and graphical descriptions for PLC programming. Control programming can be done either by the step-by-step control of sequences or by the logical linking of states.

Figure 4.2 shows a comparison of graphical sequence description and textual description using program code.

The graphical descriptions are GRAFCET, Ladder Diagrams (LD), and Function Block Diagrams (FBD). These are based on different concepts of graphical representation. GRAFCET is a simple sequential function chart based on states and other switching conditions. The representation by means of GRAFCET is particularly suitable for training purposes.

Ladder diagrams are primarily used to describe contactor circuits, but programming with ladder diagrams remains popular in practice due to its simplicity. Conversely, the functional schematic modeling concept of the function block diagram is borrowed from electrical circuitry that uses logic gates and is less common.

Programming with the instruction list happens in very close proximity to the controller, offering experienced programmers the possibility of machine-oriented optimization. However, the instruction list is considered as outdated and will be removed from the standard sometime in the future.

Textual description using structured text in a high-level description is now becoming more and more important.

Interchanging between different graphical programming languages is generally feasible, but some functions, such as counters, timers, and bit operations, may pose limitations here.

Despite the considerable degree of standardization in PLC programming, manufacturers of PLCs tend to offer language extensions and unique development environments to set themselves apart from the norm.

A range of manufacturers and product alliances offer commercial solutions tailored to different sectors of automation technology:

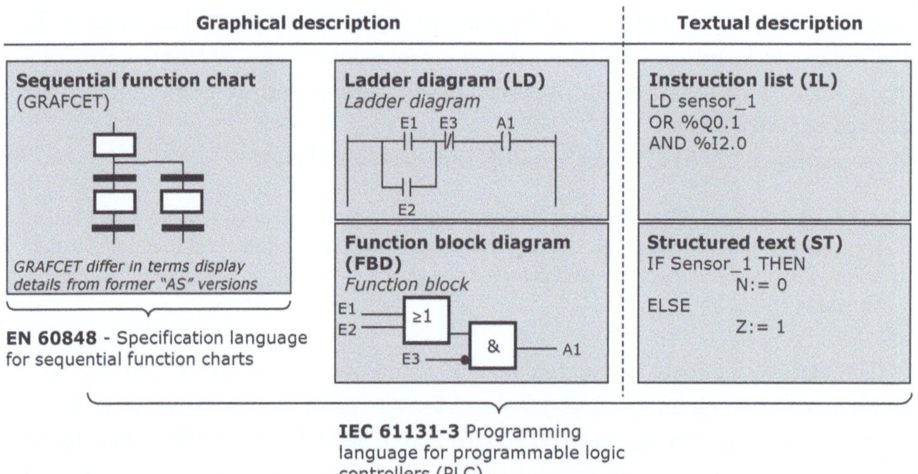

Fig. 4.2 Programming languages for programmable logic controls

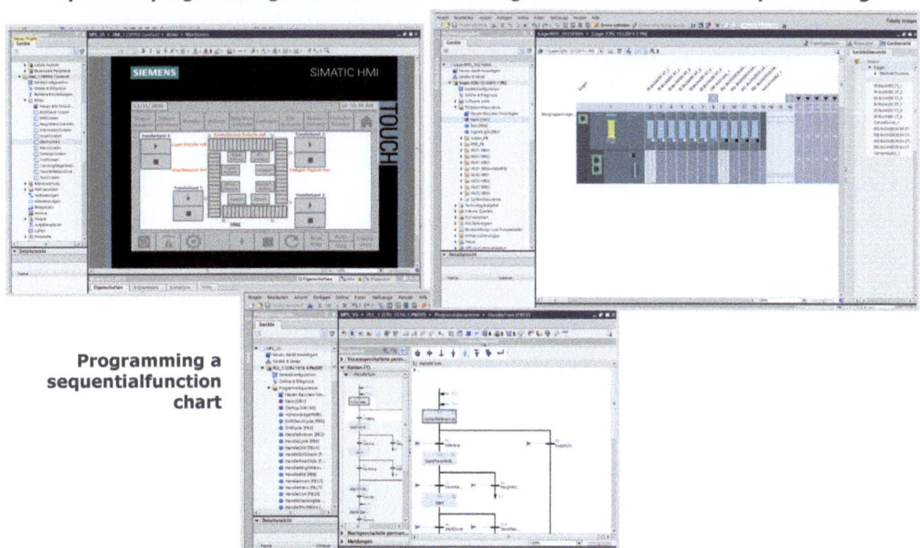

Fig. 4.3 Excerpts from a commercial programming environment (Siemens TIA portal)

– TwinCAT from Beckhoff allows the integration of Microsoft Visual Studio
– The TIA Portal from Siemens is an integration platform for controls
– The CODESYS development platform is offered as a combination from different manufacturers
– Other automation vendors offer numerous product-specific platforms.

In addition to programming control functions, commercial development platforms support the creation of human-machine interfaces for industrial automation applications. These platforms provide the extensive data acquisition, analysis, and logging capabilities needed to visualize process information in a user interface or to manage users and system components. The functionality provided by these platforms exceeds standard requirements.

Figure 4.3 illustrates the user interface of a commercial development environment that enables system configuration, programming, and the design of status indicators for a production facility.

4.2.2   Higher-Level Coordination of Processes

The coordination and supervisory control of technical processes in industrial production system automation requires interaction between automated systems and components. A supervisory control system ensures the coordination of individual automated industrial facilities or entire networks of them.

The term "supervisory control" is defined by IEC 60050-351 as

[…] an appropriately designed action on or in a process to achieve specified goals.

Operating complex cascaded automation systems requires coordination that extends far beyond the management and monitoring of individual processes. Consider for a moment the operation of large-scale process industry facilities such as chemical plants and refineries, food production (such as dairies and breweries), water supply systems, or the coordination involved in manufacturing and logistics, for example in the automotive industry.

Shifting the focus from control programming to overall system development highlights the wide range of additional tasks involved in creating software for these facilities. These tasks extend far beyond just programming control sequences. In many cases, there are regulatory contingencies for safe operation, for example in water treatment or food processing, which must be ensured and documented.

The functions of a supervisory control system are as follows:

– Integration of SCADA, MES and MIS systems to ensure overall system coordination.
– Realization of the man-machine interface for the comprehensive operation and monitoring of the system in control rooms or by means of teleoperation.
– Diagnostics for the maintenance or remote maintenance of systems and components, for example by manufacturers
– Information management of technical data for asset management, i.e. individual components and system data.

The primary role of a supervisory control system is to assist the operator. It monitors and coordinates the various stages of a process, such as start-up, regular operation or maintenance tasks. In the event of a malfunction, the system identifies the causes and the necessary actions for resolution or emergency operation. In large factories in particular, operators can comprehensively monitor and document operating conditions.

These control systems also facilitate predictive maintenance activities by using past behavior to anticipate future system responses. This early detection capability helps to identify recurring failure patterns.

Moreover, supervisory control systems manage information related to numerous system components. Parameters can be stored in a database and associated with automation components, thus helping to organize information.

These systems also support changes to existing setups. When components of a system are replaced or upgraded, control systems log the relevant functionality for component management. Change logs along with version management in automation components are critical practical considerations.

Specific industry requirements are causing IT vendors to focus on industry-specific control systems. Industries such as chemicals, food production, pharmaceuticals etc. have seen the emergence of specialized control systems offering integrated hardware and soft-ware solutions and ensuring consistent information processing and control.

The move to integrated control systems requires the development of numerous software systems, even with existing control platform options. This necessitates the use of higher-level programming languages (e.g. C#, .NET, etc.), graphical interfaces, and databases.

The market provides a variety of solutions tailored to specific customer needs which are typically implemented as client-server architectures. However, in large enterprises that monitor numerous process variables, alarms, and long-term archives, managing such large amounts of data requires further analysis or processing to maintain control.

Supervisory control and data acquisition systems are application-specific software sys-tems in which each implementation is relatively unique. Implementations range from simple management systems with a few process objects and workstations to medium-sized systems and large-scale control systems involving hundreds of thousands of process objects, servers, and workstations.

4.3   Software for the Automation of Products

The control units used in product automation consist of highly integrated components made up of one or more microchips. These controllers are mass-produced and are used in large quantities in automobiles, home appliances, and various other products. Depending on the application, product automation control units are designed to be cost-effective in order to keep the total product cost at a competitive level.

However, due to the operating environments, these product automation controllers have stringent requirements in terms of environment, reliability, and durability.

4.3.1   Development of Embedded Systems

In product automation, systems are configured from hardware components and pro-grammed using high-level programming languages. Control systems in product automa-tion involve electronic control units (ECUs) that are built directly into the device or component. This type of system is known as an "embedded system" because the controls are barely visible from the outside, i.e., they are deeply embedded in the overall automated system. Embedded systems consist of microcontrollers, which are microprocessors equipped with input and output electronics and real-time software. Different microproces-sors are used depending on the application's computational requirements. Communication in product automation control units happens through specific interfaces such as buttons, rotary switches or LCD displays integrated into the products.

Programming is done in high-level languages close to the hardware using development systems or graphical editors. Sequences critical to execution may also be developed in assembly or machine code to ensure real-time capability. Software development in this domain is characterized by the resource constraints of microelectronics, which require a deep understanding of the efficient use of resources such as memory and processing power. Other challenges for the development of embedded systems include power constraints and limited bandwidth communications.

In product automation, numerous development environments and a wide range of operating systems can be used for real-time programming.

4.3.2   Programming Languages and Development Environments

Higher-level programming languages offer a variety of programming features. Table 4.2 presents a comparison of some common high-level languages used in automation. These languages can be compared or characterized on the basis of several criteria. These include the performance and efficiency of the code, which indicate how efficiently computing resources are being used. In particular, there are significant differences between older languages such as C and newer languages such as Rust in terms of memory management mechanisms, their level of abstraction with respect to programming paradigms or so-called type-safety approaches.

User surveys allow us to draw conclusions about the use of languages: Today, C is still used intensively (approx. > 50%), followed by C++ (approx. > 20%). Java, C# and Python are used in about the same numbers, but in much smaller percentages ranging from about 2 to 6%. Assembler is also used to a similar extent. The relatively new language Rust leads a niche existence.

The **C** programming language is considered a universal "low-level" programming language with a small set of keywords and language constructs. While C does have data

Table 4.2 Comparison and evaluation of programming languages for use in real-time programming

	Executable code efficiency	Real-time behavior	Memory management	Abstraction level of the language
C	High efficiency	Yes	Simple management, risk of memory leaks	Machine-oriented with the option to include an assembler
C++	Only if partial extents are used	No, e.g. stream command and exception handling	Yes, but not deterministic	Very high
Java/ Python	No real time, efficient due to runtime optimization for dynamic actions	No, e.g. garbage collector, memory management, exception handling	Not directly, but through runtime environment	Very high
Rust	High efficiency	Yes	Special concept	High
Assembler	Very high	Yes	Difficult	Very low, costly, non-portable

types, they are not type-safe. The language permits direct memory access through pointers, which can be very flexible but can also result in complex errors. C was created at AT&T-Bell Labs between 1971 and 1974, was popularized by Brian W. Kernighan in 1978, and has since become standardized as ANSI C. It is an excellent choice for automation technology applications since the code generated offers a high execution speed and a small code size. This allows the creation of portable programs and the use of inline assemblers in close proximity to the machine. C is frequently used for real-time programming, and there is a wide range of commercial development environments and open-source solutions (including GNU-C) as well as implementations of system program libraries for real-time programming.

C++ is an object-oriented extension of the C programming language. It was introduced in 1986 by Bjarne Stroustrup.

C++ is a programming language that combines the features of C with abstract language concepts. However, this combination means that many object-oriented language concepts are only partially capable of operating in real time. Nonetheless, C++ contains a subset of C, which means that it can be used for real-time programming with some restrictions.

C++ is generally regarded as a powerful programming tool and is in widespread use. However, implementing compilers for C++ can be relatively complex. Fortunately, a wide range of commercial compilers are now available.

Java is an object-oriented language with strong typing.

Sun Microsystems developed this programming language in 1995 to create Internet applications, but its real-time capabilities are limited. Special adjustments are necessary to make it usable in automation. Nevertheless, it is widely used for higher-level operating and planning programs, and it can be accessed as a runtime environment for multiple operating systems.

Python is multi-paradigmatic, i.e. both object-oriented and functional. Python is a dynamic scripting language with a limited number of keywords. One of its outstanding features is its readability and ease of comprehension, and a wide range of runtime and development environments for various operating systems are accessible. However, if real-time automation is required, Python may not be the best option. One potential solution is to translate Python code into C or Assembler, thus enabling real-time capabilities.

Rust is a relatively new multi-paradigmatic and versatile programming language. Rust is a programming language that generates powerful and type-safe code with automatic memory management. Despite being a relatively new language, it is gaining in popularity. Rust programs are particularly advantageous in real-time programming, for example for boot loaders (startup programs), due to their efficient and high-performance code.

Assembler can be translated almost directly into machine language or machine code because the Assembler instructions and programming are extremely hardware-oriented. The resulting programs are highly efficient, providing the best possible speed and code efficiency. Assembler is often used in conjunction with C programs, where only a few critical execution segments are written in Assembler and all other program components are written in the high-level C language.

Assembler is designed specifically for one type of processor. However, this also means that assembler programs must be adapted for different processor families by means of significant effort and expertise.

Aside from the objective evaluation of a programming language, the practical considerations of manageability are also significantly weighty. It should be realized that not all programming languages or components can deliver real-time performance or produce very efficient executable code.

As a result, the following practical challenges emerge:

– Efficient verification of executable code is critical for automation applications with limited memory and computing power. Problems can arise as compilers add a so-called compiler runtime to the code. As a result, even shorter source code can produce a large program. This can result in unreasonably large programs after compilation following changes, causing problems with memory and computing power.
– To ensure optimal real-time behavior, it is essential to avoid instructions, functions, or operations with non-deterministic behavior. This includes automatic memory management, which may flush memory at unpredictable times, consuming processing time and negatively affecting real-time behavior. Exception handling should also be avoided as it can cause unknown runtime behavior when program segments are skipped due to errors. Similarly, dynamic binding and operator overloading, while useful, should be avoided as they can lead to unknown real-time compilation.
– Memory management is a critical concern, not only in terms of real-time behavior. A memory leak can cause problems in automation systems, particularly when it continuously allocates more memory than it releases. This issue is particularly common in C programs, which may crash due to a memory overflow during continuous operation.

Advanced development environments are usually available and can be used for different processor families.

The decision for or against a programming language can be based on a variety of considerations: the implementation of development platforms for certain hardware, specifications of the customer or the competences of the employees. Integrated development environments offer continuous support during software creation using standard languages.

An integrated development environment usually consists of:

– An editor with keyword completion, syntax, reference and definition checking, and versioning.
– A compiler/linker which allows source code to be compiled and, where appropriate, error messages to be formatted as direct references to source code passages,
– A debugger which allows execution to be interrupted and variable values to be read and
– A runtime environment with software libraries that help developers create applications including user interfaces.

The availability of an integrated development environment can be crucial for efficient development.

4.3.3   Operating Systems Used in Automation

Real-time operating systems are an important part of the automation software development solution.

According to DIN E 19233:

> Operating systems are defined as software that controls the execution of programs and may provide services such as resource allocation, scheduling, input-output control, and data management

Operating systems enable application programs to run independently of the underlying hardware by allowing the kernel to control the entire system. They organize input and output, manage memory for individual computing processes, and facilitate inter-process communication. In automation technology, it is important to use operating systems with real-time capabilities in order to ensure that computing processes run on time. Real-time operating systems help to abstract the specifics of the hardware in an appropriate way, using concepts of hardware abstraction via firmware and drivers to form a uniform interface for accessing the hardware. Several types of real-time operating systems are available for practical use, and they are based on different approaches in each case.

Table 4.3 provides a brief overview of some common real-time operating systems.

Table 4.3 Overview of real-time operating systems

Name of product C	Manufacturer	Processors/field of application	Type of license
QNX	BlackBerry QNX	Certified for most microprocessors (X86, ARM, PowerPC, etc.) with applications in aerospace, machine control, etc.	Commercial system
VxWorks	WindRiver Systems	Certified for common microprocessors (X86, ARM, PowerPC, etc.) with applications in aerospace, machine control, etc.	Commercial system
RTA-OSEK	ETAS GmbH	PowerPC, ARM, chipsets of other manufacturers; application esp. for AUTOSAR	Commercial system
FreeRTOS	Real Time Engineers Ltd.	Market leader with widely deployed embedded systems, supports many well-known manufacturers	Open source
Yocto	Embedded Linux Foundation	Embedded Linux for embedded and IoT software	Open source
Ubuntu	Embedded Linux Foundation	X86, PowerPC, ARM, MIPS and others	Open source

Real-time operating systems with modest memory requirements and exceptional run-time performance are readily available. Conventional operating systems also provide add-ons for real-time operation.

The criteria for choosing a real-time operating system are diverse, ranging from technical specifications such as kernel size, processing velocity or certain feature sets (such as scheduling) to factors like ease of learning and cost-effectiveness. Product liability concerns are a significant consideration, particularly when comparing open-source software to commercial products.

4.3.4   Automated Software Development Processes in Product Automation

In product automation, sophisticated tools such as program generators, high-level programming languages, real-time operating systems, and software frameworks with multiple libraries are commonplace. Nowadays, manual programming of source codes is rarely utilized as codes are predominantly generated and merged automatically based on models. Large teams in automotive software development, for example, utilize customized tools to generate software in an industrial development process. While comparable development setups exist in industrial automation, product automation is distinguished by far more complicated software systems.

Software systems (software stacks) in product automation can be very large and may include many interconnected components.

> A software stack is a set of software systems that work together in a well-defined, orderly fashion to achieve a specific result. A software stack may include definitions for interfaces and programs.

The use of interactive program generators causes a shift from detailed development in the programming sense to development management in a so-called toolchain.

Figure 4.4 sketches the various stages of an industrial software development process, including specification, model-based design, configuration and version management of the source code, and the automatic integration of the generated code into a control system (so called "flashing").

The diagram depicts the various development tools utilized in a development chain, starting with the specification and progressing to the creation of source code with the help of models and programs. To guide industrial software production on the basis of experience, software engineering process models are employed.

Modern software development typically utilizes pre-existing operating systems, libraries, and other software structures. Third-party software components are crucial for the final outcome and are integrated during the make/build phase to be uploaded, or flashed, into the ECU. However, thorough testing is required to ensure the reliability of software of

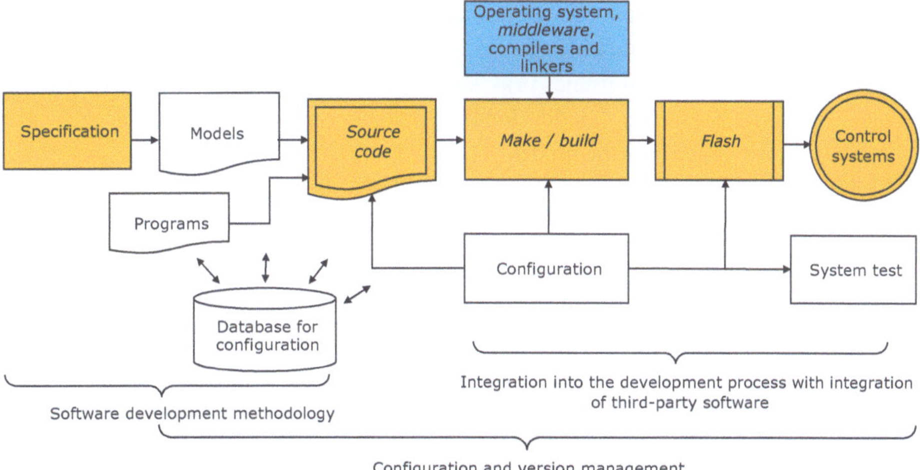

Fig. 4.4 Software development with an automatic tool chain

this kind, which includes system testing and precise definition as part of the development process.

However, this process does not form a program but a combination of different software configurations resulting from the development phase, the components delivered, and the requirements of various deployment environments, among other factors. The management of these configurations is crucial.

An industrial tool chain for software creation consists of the following main components:

- Tools for software engineering, i.e. the systematic generation of requirements,
- Models of systems or subsystems which can be used to simulate the technical process and model controls.
- The source code, which today is often designed using graphical development interfaces or generated from models using code generators.
- Software management and configuration. This is very important because, in automation technology, many versions and variants of software can be in use at the same time.
- Software testing. This is becoming increasingly important in automation because the correct system behavior can be validated and verified by a test run in simulation and in real operation.

The use of automatic end-to-end software development with seamless solution modules offers advantages since the resulting source code has no or few errors and functions in a more mature way.

In order to implement toolchains for development, a number of prerequisites have to be met:

- The controller models should be defined so that individual development focuses mainly on configuring and coordinating them.
- The operating system and frameworks should be fixed in module libraries so that they can be reused. This means that development partners, vendors of software components etc. can be integrated into the process.
- The microelectronic hardware platform should also be defined, since it can be very costly to change the software production pipeline with compilers, special libraries, etc.

Maintaining toolchains with various development tools from system vendors in their current and operational state can be demanding, given the practical use of numerous vendor products and tools. Continuous changes can occur due to their ongoing evolution, even if the actual control program remains unchanged.

However, the use of automated software development chains does not only offer advantages. The process chains are monolithic and difficult to change in their overall configuration, leading to a loss of development agility. If there is a need for an ad hoc change in the development process chain, the use of other IT tools can have a non-transparent impact on software quality. The mere fear of such difficulties can prevent changes from being made.

4.4   Software for Cyber-Physical Automation

The realization of cyber-physical automation systems has advanced significantly. Today, there are many ways to develop applications (apps) based on web service platforms and process large amounts of data on servers in the cloud.

For implementation purposes, it is necessary to clarify a few questions regarding the IT system architecture:

- How do the individual control units obtain information about each other so that they can exchange information?
- Which tasks need to be centralized and which need to be decentralized?
- Where is the data stored and how is it exchanged?
- How can continuous integration and delivery (CI/CD) be implemented?

Because of the complexity of the implementations, the basic functionalities are almost always implemented in conjunction with one or more IT platforms.

Implementations are often highly sophisticated due to system characteristics, with basic functionality almost always being realized in conjunction with one or more IT platforms. Table 4.4 lists the functionalities required for the management and operation of cyber-physical automation systems. The table indicates the level of complexity of the structure. Not all functionalities are mandatory here; the attention can be focused on the implementation of only a few functions if necessary.

Table 4.4 Functions of software for the realization of cyber-physical automation systems

Functionality	Description
Device management	Activation and control of individual distributed components
Identity and access control	To decide on an entity's integration, rights, roles, and relationships, the entity must be defined and identifiable in the system.
Process management	Specifies the form in which data and information are to be exchanged between the individual entities.
Digital twin / virtual entities	The models and data lakes of the systems can be managed in a distributed manner if required.
Analytics	The data is analyzed using various methods to generate information.
Applications (apps)	For instance, mobile applications require appropriate software interfaces.
IT service organization	The execution of individual services requires costly implementation on a platform.
CI/CD functionalities	Centralized software management to automate updates and rollout

Managing individual participants, especially in large networked systems with data spaces, is challenging when it comes to activating and controlling devices. In IoT systems, it is assumed that new devices are constantly being added or removed from the system. However, subscriber management must remain reliable and secure despite these fluctuations.

Identity management is critical. Both individual devices and their users must be authenticatable, meaning they should be able to verify their "authenticity" via a certificate or another means before gaining access.

Even with robust security measures, the possibility of data leakage always exists when it is stored as virtual entities or as a digital twin in a centralized location. Hence, the choice of data storage is a critical decision that requires careful consideration as to whether centralized or decentralized storage is the more viable option. The storage method selected can significantly impact how data is assessed and managed.

Access to IT services requires applications built on an application platform. These platforms rely on an IT infrastructure to execute IT services and, if necessary, leverage additional services for processing and analysis.

Processes such as CI/CD can also be used to roll out software updates automatically across the system, thus ensuring that the software is always up to date. Today, only a few vendors, some of whom have formed alliances, offer a comprehensive IT infrastructure for the automation of Internet of Things systems. These few vendors currently hold a dominant market position due to the significant development effort required for in-house core components.

The ability to implement cyber-physical automated systems is increasingly dependent on the use of cloud software from very large IT companies. While there are drawbacks to this dependency, these IT infrastructures allow better implementation for corporations with global operations. Centralization through highly interconnected automation systems offers the opportunity for global standardization and uniformity, leading to cost advantages.

## 4.5	Control Paradigms and Service Architectures

There are various different paradigms for controlling automation systems. This chapter covers cyclic and event-driven control as well as service architectures.

### 4.5.1	Basic Control Paradigms of Cyclic and Event-Oriented Control

A difference is made between cyclic and event-driven control, each of which has different characteristics, advantages, and disadvantages.

- In cyclic control, polling is periodic, time-controlled, repetitive, and largely centralized. In this case, communication takes place at repeated intervals and in the same sequence, so that the nodes are regularly polled and then coordinated accordingly.
- In event-driven control, the individual participants can be more decentralized and distributed. In this case, communication is not predetermined, but the participants come together as needed and communicate only when events require them to do so.

We can observe the various control paradigms at play if we consider a traffic light intersection as an example. The technical system involves cars and drivers navigating the intersection while being guided by the traffic lights and other vehicles.

The cyclic control corresponds to the rolling traffic light sequence that periodically generates a "stop-and-go" traffic flow in sync with the switching cycle of the traffic lights.

However, if an emergency vehicle approaches with flashing lights, an event-driven paradigm takes over, overriding the cyclic control to prioritize the emergency vehicle's passage, even if the traffic light is green.

Under normal circumstances, time-based cyclic control is employed due to the alternating traffic signals. However, in exceptional situations such as the approach of an emergency vehicle with flashing blue lights, an event-driven paradigm which overrides the time-based control is used.

The implementation is relatively straightforward: imagine a control system that manages traffic lights in a time-based rhythm while also responding to events such as the arrival of an emergency vehicle. In this sense, a clock generator can generate cyclic events while simultaneously handling external events.

This example illustrates the practical value of combining both paradigms, a procedure which is relatively common.

4.5.1.1 Cyclic Control (Also Known as Clocked Control)

The concept of cyclic control involves querying based on a time grid, meaning there are equidistant intervals at which processing occurs. Cyclic control is therefore easy to predict. In line with the spirit of DIN IEC 60050-351 (2013), it can be defined as follows:

> A **cyclic or clocked control** is a type of open-loop control that works with a clock signal in which changes of output variables are trigged by changes of the input variables, but only when the clock signal allows these changes.

There is only one central computing process that checks logical connections. Control occurs synchronously with a cycle that triggers the processing of inputs and outputs in a fixed timed grid. This means that output signals are based on a logical combination of input signals.

An isochronous cycle with equally spaced intervals ensures timeliness when properly designed. Time control is triggered by the passage of time, for instance when a sensor value is transmitted.

Processing with cyclic querying is also known as cyclic execution or polling. Its computation times are easy to estimate, and the logical connections between output and input signals are straightforward. Due to its simplicity, cyclic control is often used to manage process flows.

The basic concept of cyclic control is shown in Fig. 4.5. The control programs are executed by processing the instructions in constant cycles.

The basic concept of cyclic control is illustrated in Fig. 4.5. Control programs are executed by processing instructions at fixed intervals. The control operates within a defined time grid T_Z. If a change in the process signals occurs at time t_{Event}, it might take a while (T_E) for this event to be recognized in the control, since the next cycle must occur. Processing by the control program requires a processing time (T_V). Usually, the clock frequency of the microprocessors in the controller is considerably higher than T_Z, so that

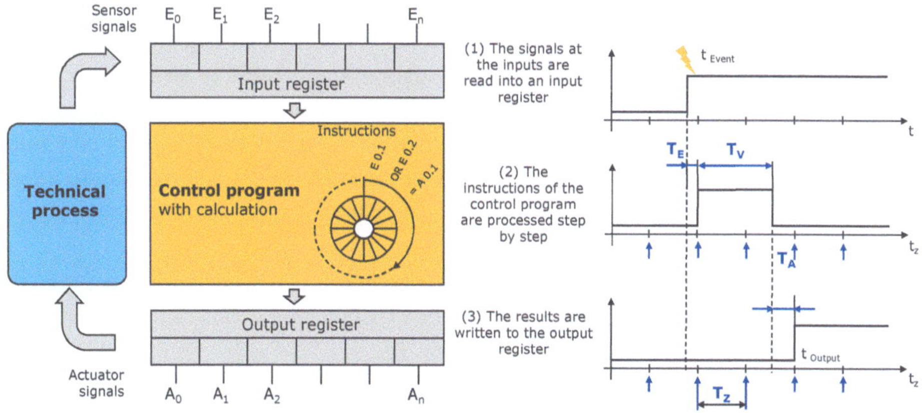

Fig. 4.5 Realization of cyclic control

the processing of the control program can be considered as quasi-continuous computation.

Finally, the computation of the output values is completed after a time T_A, and the actuator signals are introduced into the technical process at the control output via the actuators at a time T_A. Depending on the event occurrence and processing time due to cyclic processing, the maximum response time can be almost twice the cycle time. This is because an event has to be identified first. In the worst case, this happens immediately after the inputs are read, so almost a full cycle elapses before the calculation takes place within the next cycle (T_Z). Thus, almost two cycles pass before the event is addressed. In addition, there may be delays in sensor response, transmission times between sensors and actuators, and delays in the actuators.

Cyclic programming is a popular method for performing time-repeating automation tasks. Programmable logic controllers (PLCs) operate on the principle of cyclic programming and are widely used in practical applications.

While cyclic control offers several advantages such as easy program planning and modification without complex organization programs, it also has a systematic disadvantage. Each cycle involves continuous communication with sensors and actuators followed by computation, which constantly consumes communication bandwidth and energy. For this reason, cyclic control is less suitable for battery-powered systems that aim to minimize energy consumption.

4.5.1.2 Event-Driven Control

Event-driven control responds to occurrences in the process and is defined according to DIN IEC 60050-351 (2013):

> (**Event driven control** is a type of non-clocked open-loop control that works without a clock signal and in which changes of output variables are only induced by changes of the input variables

Event-driven control is also referred to as signal-based control since the occurrence of events manifests itself in the form of changes in input variables, i.e. signals.

When an event such as the activation of an alarm, the initiation of an emergency stop function, or the shutdown of a motor due to the triggering of a limit switch takes place within the technical process, a corresponding routine is started in the control computer. This routine is responsible for performing the necessary actions in response to the event. In this way, events trigger the invocation of a control program tasked with processing the specific event.

Figure 4.6 depicts the timing of an event-driven control. The response time between the occurrence of the event t_{Event} and the appearance of a potential action at the output at the time t_{Output} depends solely on the processing time T_V.

Events can also be temporal conditions, i.e., the unfolding of predefined time intervals such as a specification that certain process variables should be checked every 10 min.

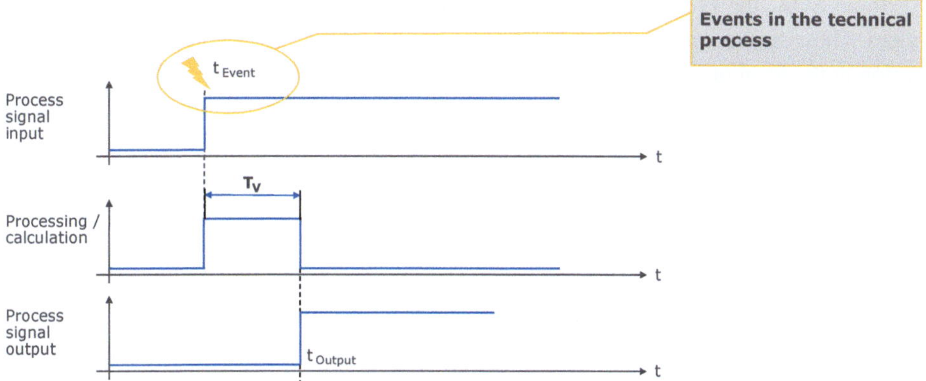

Fig. 4.6 Example of an event-driven control timing sequence

Thus, in the event-driven control paradigm, temporal sequences unfold during operation as a result of events. In addition, user instructions received by the control system can be interpreted as events requiring a system response. It is also possible for a lower-priority event to be interrupted by one with a higher priority.

The advantages of event-driven control are obvious: it demonstrates significant flexibility in handling unforeseen events by dynamically responding to service execution on the basis of events and temporal conditions. Events are processed in parallel, resulting in convenient and efficient execution. In particular, event-driven control optimizes energy consumption and communication by computing and communicating only when necessary.

However, one drawback of this is the uncertainty in the order of service execution as the processing time of lower-priority processes can be delayed by events with a higher priority. The resulting execution sequences can become complex, potentially leading to a temporary loss of real-time capability if lower-priority processes are neglected.

In addition, dealing with a large number of events complicates traceability, creating an impression of "deterministic chaos".

4.5.2   Service Architectures

In service architectures, IT components are broken down into modularized services that provide access to system resources. To request a system feature, users make service calls. Services are information technology services that can be executed when called. This allows for the combination and orchestration of appropriate service bundles.

There is a clear difference between a service-oriented architecture (SOA) and a microservices architecture.

Both IT architectures handle individual services that can also be used for control purposes. An incoming event such as a cyclic call or an event-triggered request initiates a service call.

In an SOA, the IT architecture consists of multiple software layers that provide services. These services interact across different layers and may use a complex protocol definition in the form of a "service bus".

The microservices architecture is characterized by simple, well-defined, and independent services that can be executed individually.

The difference between SOA and a microservices architecture lies in the number and granularity of the individual services. In a microservices architecture, each service represents a specific function, whereas an SOA can include much more complex services. Another difference is the complexity of communication between services: Microservices tend to have simpler structures, while SOA can involve complex protocols.

Both IT architectures promote modularity, scalability, and maintainability through their structures.

Service-oriented architectures (SOA) have become very popular in a wide variety of areas and are being used extensively not only in automation technology, but in other areas as well.

The implementation of SOA and microservices architectures differs depending on the application area and can lead to powerful software systems. The number and complexity of services, the specification of protocols for interaction between services, and the structure of architectural layers play a significant role here.

A very basic structure of an SOA is shown in Fig. 4.7:

– One of the central elements is the management of services in a registry in order to find their origin, rights, and other conditions, and to make them discoverable.
– To ensure that the service-oriented architecture can be extended later, all service interfaces should be uniformly defined.

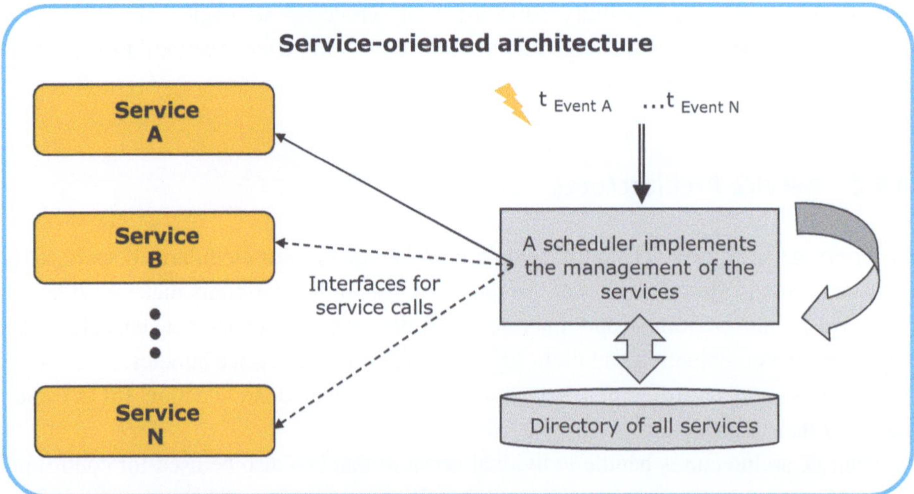

Fig. 4.7 Concept of service-oriented architecture (SOA)

– Schedulers are used to realize the quasi-simultaneous execution of services invoked. Scheduling involves the generation of a temporal occupancy plan to allow the easy location and optimal allocation of individual IT services or their resources.

Individual IT services or their resources should be easily located and optimally allocated. For instance, if several control algorithms are to be executed quasi-simultaneously, the required services must be started on time and run alternately. Comprehensive descriptions can be found in Brown et al. (2012).

Both the service invocation procedure and the scheduling method play an important role. To ensure that all deadlines are met on time, priorities can be assigned.

It is evident that reliable service invocation is of central importance. If service invocations are not communicated or timed properly, the reliability of the service-oriented architecture-based control is compromised, leading to obscure failures that are difficult to identify and trace.

4.5.3   Agent-Oriented Architectures

Another architecture is based on so-called agent systems. This abstracts the previously introduced concept of service architecture to form a concept of agents.

According to VDI Guideline 2653 Sheet 1 (2018):

> [An] agent is an encapsulated (hardware/software) entity with specified objectives. An agent endeavors to reach these objectives through its autonomous behavior and by interacting with its environment and other agents.

Agents are thus attributed with far-reaching properties of autonomous systems. They have the capability to provide required performance according to their abilities. Agents can decide whether this performance should be executed fully, partially, or not at all. Hence, agents have the ability to act autonomously within their designated objectives, roles, and spaces of action. An agent's behavior depends on its perception of the environment. On this basis, agents utilize their capabilities to influence the environment by means of actions. For this to happen, agents need to process sensor information, make decisions, perform actions, and interact with other agents, for instance in order to negotiate with them in terms of achieving goals, and potentially work in a collaborative way.

Research studies on agents deal with how typical use cases in automation technology appear and whether and how specific patterns (see VDI Guideline 2652 Page 3 [2018] and Page 4 [2022]) can be delineated. This involves examining their deployment in various practical systems such as modular production facilities and material flow systems or their use in engineering. It also encompasses the description of typical patterns in terms of agent topologies and roles of agents.

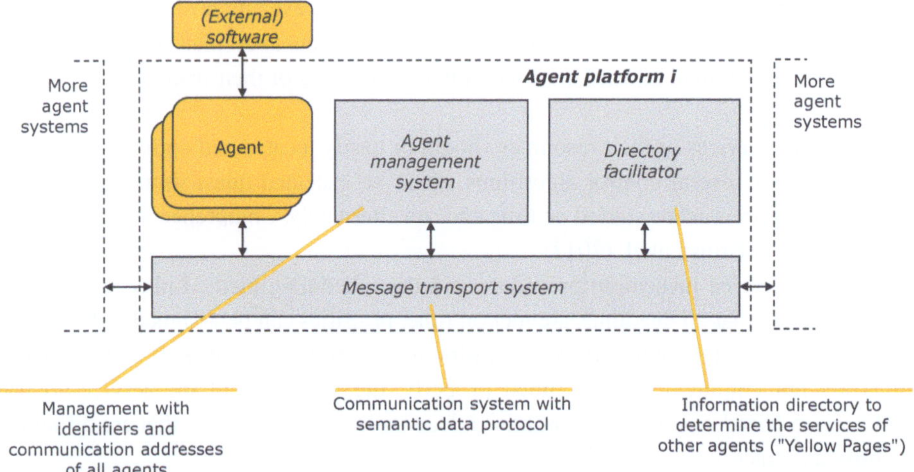

Fig. 4.8 Reference architecture for agent systems (adapted according to FIPA, 2004)

Frameworks, programming languages etc., have been implemented and made available in this context (see VDI Guideline 2652 Part 3 [2018]).

As early as in FIPA (2004), a reference architecture for agent systems and agent interaction was defined from which software frameworks emerged.

The FIPA reference architecture (see Fig. 4.8) specifies a system for message transmission, governing communication between agents and defining, through the JADE language definition, a type of vocabulary of a language for communication. Additionally, the reference architecture governs two information services through which agents can be managed in terms of software and found via a service directory. Also, an agent management system enabling the management of agents is specified.

This agent architecture can also serve as a model for much simpler service-oriented architectures or microarchitectures. However, agent-oriented architectures pursue much more ambitious services, aiming for system autonomy in perception, understanding, and action (see also Chap. 2, Fig. 2.7).

While the definition of agents offers a broad framework for designing autonomous units that solve problems independently or collaboratively, the way in which perception, understanding, and action are realized is left to the developers. Consequently, recent research has been diverse, focusing on how agent topologies should be structured and how interaction and communication between agents can be implemented.

Although negotiation patterns of agents based on deterministic control systems have been researched well so far, as summarized by Vogel-Heuser et al. (2020) for the implementation of Industrie 4.0 concepts and demonstrated by prototypes, the way in which agents actually implement the abilities of perception, understanding, and action largely remains an unanswered question. How artificial intelligence can ensure goal attainment of this kind through intelligent behavior has not yet been clarified and will require further research.

Nevertheless, some interesting aspects of the interaction between agents and agent systems have already emerged in the field of automation technology, and numerous examples demonstrate the successful implementation of agent systems of this kind.

4.6   Prompts for Reflection

This chapter provides an overview of the industrial software landscape for use in and with automation systems. While it provides a concise introduction to the specifics of developing software for real-time applications in automation, its aim is to equip the reader with the ability to explain basic concepts and describe the specifics of software for automation in different industries according to the defined categories.

It becomes evident that the methodologies used in these industries differ significantly, often reflecting the skill levels of the people involved and the different contextual frameworks. Successful implementation of software projects requires specialization and deep immersion in development platforms which is sometimes specific to the processes established within the respective organizations. As a result, a deep understanding of technical details is essential. Online training courses offered by major vendors or open-source development environments can be instrumental in this regard.

Given the dynamic nature of the field of automation software, achieving comprehensive mastery is challenging. Nonetheless, the following questions offer an opportunity to explore the subject in depth:

What software platforms are available in industrial production system automation?

The programming of control and supervisory systems in production system automation has long relied on programming environments provided by established vendors of automation components.

Conduct Internet research to determine the following:

a. Which industries are offered specific control and supervisory systems?
b. Which control system development platforms are being promoted in Europe, America, and Asia?

How do you rate the automated development of software through the use of development chains in product automation?

In the field of product automation, the creation of software for embedded systems is becoming increasingly complex due to the use of numerous programming languages and development environments integrated into extremely complex tool chains.

a. What are the advantages and disadvantages of creating software using automated development processes versus creating software using integrated development environments?

b. How do you assess the agility of development via toolchains compared to the resilience provided by the use of mature software components?

How do you evaluate the impact of entirely new technologies on cyber-physical automation systems?

a. What role can cloud-based IT platforms play in the evolution of automation technology?
b. Can industry-specific control system platforms be replaced by cyber-physical automation system approaches in an Internet of Things in the future?

How do you view the challenges of developing service architectures?

The development of SOA and microservice architectures in the field of command and control is now in progress. Architectures of this kind used for platform-independent data exchange and as a basis for system integration are gaining in importance. Conversely, there may be a tendency toward simpler in-house developments.

a. What are the advantages and disadvantages of complex service architectures compared to straightforward and "flat" in-house developments?

Further Reading

Burns, B.: **Designing distributed systems: Patterns and paradigms for scalable, reliable services.** O'Reilly, 2018. https://dl.acm.org/doi/book/10.5555/3235491

Elfatatry, A.: **Dealing with change: components versus services.** Communications of the ACM, 50 (8), 2007. https://doi.org/10.1145/1278201.1278203

Küfen, J.; Hudecek, J.; Eckstein, J.: **Automotive service oriented system architecture—ein neues Architekturkonzept und sein Potential für zukünftige Fahrzeugsysteme.** Automotive meets Electronics, 2014

Lee, Ed.; Seshia, S. A.: **Introduction to embedded systems - A cyber-physical systems approach.** Second Edition, MIT Press, 2017

Tanenbaum, A. S.; Bos, H.: **Modern operating systems.** Pearson, 2014

References

Brown, P.; Estefan, J. A.; Laskey, K.; McCabe, F. G.; Thornton, D. (Hrsg.): **Reference architecture foundation for Service Oriented Architecture.** Committee Specification, Version 1.0, OASIS, 2012. http://docs.oasis-open.org/soa-rm/soa-ra/v1.0/cs01/soa-ra-v1.0-cs01.pdf

DIN EN 60848 / IEC 60848: **GRAFCET specification language for sequential function charts.** Beuth-Verlag, 2013. https://doi.org/10.31030/2248266

DIN EN 61131 / IEC 61131: **Programmable controllers - Part 1: General information.** Beuth-Verlag, 2003. https://doi.org/10.31030/9537680

DIN IEC 60050-351 /IEC 60050-351: **International electrotechnical vocabulary - Part 351: Control technology.** Beuth-Verlag, 2013. https://doi.org/10.31030/2159569

Vogel-Heuser, B.; Seitz, M.; Cruz, S.; Alberto, L.; Gehlhoff, F.; Dogan, A.; Fay, A.: **Multi-agent systems to enable Industry 4.0**. at - Automatisierungstechnik, vol. 68, no. 6, 2020, pp. 445–458. https://doi.org/10.1515/auto-2020-0004

FIPA: **Agent management specification** Alameda, USA, Foundation for Intelligent Physical Agents, 2004. http://www.fipa.org/specs/fipa00023/SC00023K.html

VDI/VDE 2653: **Multi-agent systems in industrial automation**. Sheet 1: Fundamentals, Sheet 2: Development, Sheet 3: Application, Sheet 4: Selected patterns for the field level and energy systems. Beuth-Verlag, 2018, 2020, 2022

IT for Networking and Communication in Automation

5

Abstract

This chapter discusses the networking capabilities of information technology for automation systems. It makes a distinction between "small-scale" local networks and "large-scale" networks based on the Internet of Things (IoT).

After assessing and analyzing the requirements, we examine the key technologies for IT networking in automation technology for systems and subsystems. Due to the wide range of implementation options, the question arises as to which technologies are essential for this purpose.

- Which wired communication networks are currently deployed in the industry?
- How can wireless communication be realized and what are the possibilities for object identification?
- What is a reference model for IoT networking?
- Which IoT protocols are suitable for highly networked communication?

This chapter provides an overview of the various methods of networking and communication and their application areas in each case.

5.1   Introduction

Today, networking capabilities such as Ethernet, WLAN, or mobile data exchange have become state-of-the-art systems indispensable in our daily lives. The first standards for real-time applications in automation were introduced more than 40 years ago. One example is the CAN bus, which was introduced in the early 1980s and is still very popular today.

The proliferation of Ethernet or software system interfacing methods has introduced many additional concepts to the field. A wide range of different communication solutions allow components and objects to be connected.

Similarly, coupling options emerge with their respective standards in communication networks that connect software systems within an enterprise. In systems in which many devices communicate with each other, Internet and mobile communication technologies are spreading rapidly. In recent years, new communication standards have emerged in these areas and are now finding their way into automation. One example of an Internet standard is the IPv6 protocol along with telecommunications standards such as 5G.

It is worth pointing out that, despite the efforts at standardization, there is currently no unified landscape of networking options. Instead, the current situation can be compared to the "Tower of Babel" analogy, with a multitude of different options for realizing communication channels and protocols.

5.1.1   Connectivity and Interoperability

In the field of automation, connectivity is defined as follows:

> Connectivity refers to the ability to establish an information technology-based connection between individual components of the technical process, the automation system, and human entities.

There are at least three levels of connectivity or networking between systems and components:

- Point solutions in which individual automation systems communicate directly with each other without the involvement of additional systems. Network partners are closely coupled and use specialized communication technologies.
- Larger networks with many communication partners in which automation systems communicate within a closed system. Examples include manufacturing facilities in which different subsystems, machines, and equipment can communicate across different locations as needed.
- Extensive networks such as components of transportation infrastructure in which numerous participants such as vehicles, charging stations, and parking facilities com-

municate to enable overall coordination. Networks of this kind may also be publicly accessible and accommodate changing participants.

A comparison of these three types of communication networks reveals the diversity of technical solutions in practice.

In terms of the interconnection of automated systems, no matter whether in point-to-point or closed networks, it is feasible to network them if the systems perform relatively simple functions.

For more complex functions, and despite the availability of numerous standards in automation technology, many questions remain regarding the exchange of data and information. Networking with the aim of connecting devices and enabling comprehensive data analysis and processing is labor-intensive and requires standardization to ensure interoperability.

According to VDI 2019,

> interoperability is defined as the ability for active, purposeful collaboration between different components, systems, techniques, or organizations.

On this basis, communication needs to be standardized using technologies developed for distributed IT and software systems. Because of automation's real-time, bandwidth, and reliability requirements, Internet standards cannot replace existing communication systems such as fieldbuses.

Interoperability requires a unified architecture for integrating devices seamlessly without the necessity for manual adjustments. Data management—the exchange and management of process or system data—remains an important aspect in this context. As a result, it is clear that seeking a comprehensive standard that ensures the connectivity and interoperability of the application segments of automation technology is a broad field with significant ongoing requirements in terms of research and development.

5.1.2 Requirements for Networking in Automation

In automation practice, applications place different demands on networking. Factors such as real-time capability, robustness or resilience, and transmission speed are important in communication. The way in which components and systems are connected can also be relevant and variable, especially when systems need to communicate reliably in terms of machine safety, such as in potentially explosive environments. In addition, ease of use by maintenance personnel plays a key role in industrial production. Considerations include the provision of simple and durable connectivity that can be used with protective clothing and gloves.

Table 5.1 Selection criteria for industrial communication technology

Function and scope of services	Cost factors	Security and resilience
Real-time requirements and transmission rates	System layout	Influence caused by the disconnection of connections
Guaranteed transmission bandwidths	Granularity	Effect of the malfunction of central or end devices
Interoperability with other systems	Expandability	Troubleshooting expenditure
Energy consumption and energy efficiency	Complexity of the systems	IT security against attacks
		Functional reliability of machines and devices

Increasing emphasis is given to IT security, especially protection against IT attacks (hackers), as automation systems are vulnerable to potential breaches which could pose a hazard under severe conditions.

Current discrepancies between individual systems exist primarily in the following areas

– Various signal, interface, and protocol standards
– Collection, recording, and aggregation of information sources
– Integration of multiple heterogeneous automation components from different vendors

Table 5.1 provides an overview of the selection criteria for industrial communication technology. It delineates the scope of functionality and performance while also considering cost factors, especially in cabling. Criteria such as security and resilience describe the ability to cope with various influencing factors while ensuring proper functionality.

Bandwidth availability and timely processing are critical for real-time applications. Significant disruptions can occur if the connection is compromised due to maintenance, for example.

Energy consumption can also be critical to the application. While it plays a role in communication within industrial facilities, this role is significantly less important than, for example, the battery operation of wireless switches in mobile systems. As a result, the issue of power consumption, and consequently of battery life, is of greater concern for wireless switches.

5.1.3 Networking of Systems and Components on a Large and Small Scale

There are a variety of application scenarios for the networking of automation systems and their components. In practice, both production systems and product automation are locally networked.

The connection between so-called field devices—i.e. automation components within a technical system—is often described as "field communication". Field devices can be actuators or sensors connected to each other or to controllers.

Vehicles also integrate numerous sensors, actuators, and controllers that communicate with each other. Another example is building automation, which requires the networking of individual components. The wide range of communication methods reflects the wide range of applications.

This spectrum of different techniques for networking information technology expands even further if we consider the numerous specialized developments of the past few decades. Various communication systems have been developed during this time. Some of these systems are still in use today due to their long service life.

When communication occurs between automation devices on the same level, it is called **horizontal communication**. It involves the interconnection of machine controls, subsystems of manufacturing equipment, logistics equipment etc. by exchanging data and information via the interfaces of these system levels.

Vertical communication, in the context of the automation pyramid, means connecting different levels such as enterprise IT to supervisory control and data acquisition (SCADA) systems or directly to controllers, sensors, and actuators. Data and information at each level must be interpreted in context and normalized if processing is to meet the different information content requirements of these levels.

For example, the detailed time sequence of sensor data in an assembly machine may not be as relevant for an enterprise-level commercial IT system as the number and timing of parts passing quality inspection. However, information from the proximity network may be relevant for other partners beyond that immediate area. These include logistics networks handling raw materials or remote machine maintenance conducted by machine vendors.

For example, the connection to fleet management in the vicinity requires an edge network whereas the connection to a machine manufacturer's remote control room can be made over the Internet.

Figure 5.1 illustrates communication partners in the local network—known as the proximity network—in which automation systems and components are located in close proximity within a local field network. The connection to the global wide-area network—i.e. the Internet, where the cloud resides, is facilitated by the edge, which represents the connection point or boundary. Here, local servers enable decentralized data storage at the "edge" of the Internet.

The following terminology is occasionally used in this context. The proximity network, a communications network in close proximity to the technical process, is connected to the Internet and the cloud via edge computing. Metaphorically, the decentralized data processing at the edge is referred to as "fog": Automation system and component data is processed on various local servers (fog) before being sent to the cloud over the Internet.

Edge-cloud integration poses several challenges for automation:

– Multiple signal, interface, and protocol standards
– Multiple standards for communication in automation
– Integration of heterogeneous automation components from different vendors
– Capturing, recording, and aggregating information sources

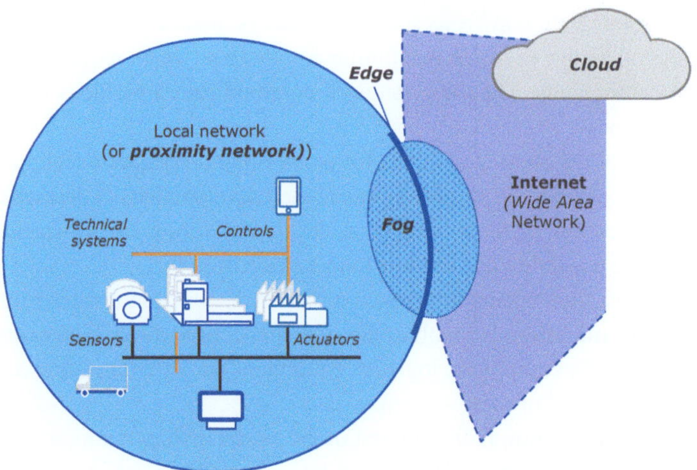

Fig. 5.1 Proximity and edge network with access to the internet and the cloud

As a result, automation data and information exchange systems often have limited mutual understanding and are typically not integrated or interoperable in practice. Despite standardization efforts, there is currently no uniform networking solution.

Communication partners can be wired or wireless. In industries in which automation devices are often already wired, wired connections are preferred due to their robustness and their lower susceptibility to interference in harsh operating environments. However, when it is necessary to access remote or hard-to-reach locations, wireless networks offer distinct advantages because they are easier to install, maintain, and expand.

5.2   Wired Systems for Local Area Networking

Wired networks are widely used today due to their robustness and availability. The technology has matured over the years and is available in many standardized variants. It is the preferred method in harsh industrial environments where safety or explosion risk factors must be considered. In industrial environments, the cost of cable installation is often not a significant factor.

However, wired systems are static in their configuration and dependent on a cabling layout that is costly to change. Also, wiring can require sophisticated infrastructures and topologies. Figure 5.2 illustrates the market shares of various wired communication systems, distinguishing between the long-standing and widely used fieldbuses and the more recent and expanding Industrial Ethernet.

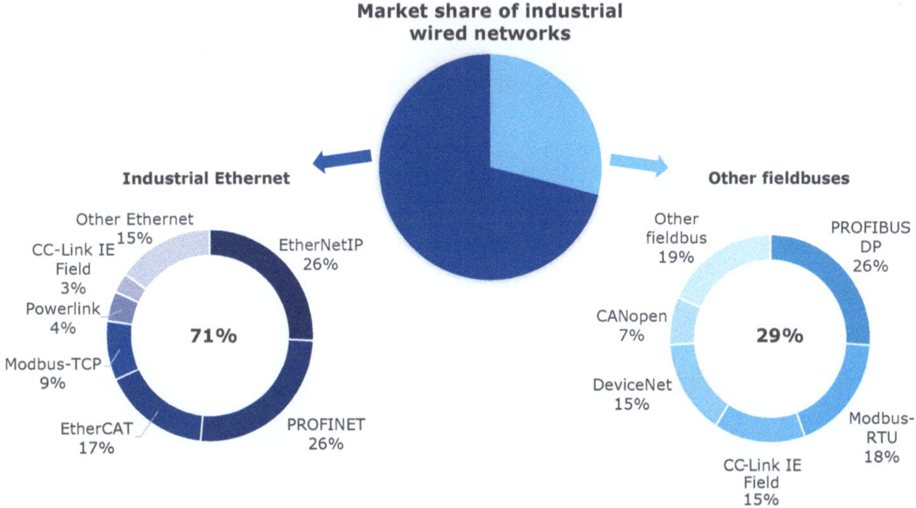

Fig. 5.2 Market share of industrial wired networks (according to a survey by HMS Industrial Networks 2022)

5.2.1   Field Communication

Field communication systems—commonly referred to as fieldbus systems—are widely used in the field. Their origins can be traced back to the standardization efforts of many decades ago. The early field communication systems of the 1980s were developed due to the lack of alternative communication technologies and are often highly application-oriented, having been designed for specific use cases. According to DIN IEC 61159 (2020):

> A fieldbus system is a system of serial data communication for data exchange in the field with specific requirements.

It is worth noting that the term "fieldbus" is frequently used as a general term for field communication systems, even though they are not necessarily wired on the basis of a bus topology. Accordingly, as DIN EN IEC 61159 (2020) further elaborates:

> A fieldbus is understood to be a functional unit for the transmission of data between several participants, where participants do not contribute to the forwarding of data transmitted between other participants.

Field communication replaces parallel wiring with serial wiring using digital transmission, thus significantly reducing cable runs between sensors, actuators, and controllers in comparison with direct wiring.

There are a number user organizations that specify and develop field communication standards. These systems are often cost-effective, tailored to meet specific needs, and

Table 5.2 Field communication systems in industrial use

Name	Area of application	Characteristic	Wiring	Reaction time	Extension (per segment)	No. of participants
Profibus	Field level of production	Multiple variants	Line	Medium	100 m (to 1200 m)	126
MODBUS	Basic field communication	Master-slave bus system	Line	Medium	15 m (customized up to 1200 m)	32
CANopen	Field level in machines and vehicles	Small quantities of data	Line	Medium	100 m (to 1000 m)	127
FlexRay	Fault-tolerant system (failsafe)	Application in vehicles	Line Star	Very fast	24 m	15
Interbus	Factory field level	Can be segmented	Ring	Very fast	Max. 400 m	256
AS Interface	Sensor/actor level	For binary inputs/outputs at low cost	Line Star Tree	Fast	100 m	62
IO-link	Robust sensor/ actuator communication	Master-slave	Point-to-point	Fast	20 m	128 per master

therefore easy to implement. In turn, a wide range of microelectronic components support these communication systems, thus simplifying and streamlining their implementation.

Field communication systems are tailored to specific demands and are hence only used for specific tasks such as the fast and secure transmission of small amounts of data.

Table 5.2 provides a brief overview of some common field communication systems.

Field communication systems are considered as simple, cost-effective, and reliable. On the other hand, Ethernet-based communication technologies are increasingly finding their way into industry. However, these methods require modification in order to meet the real-time demands of automation.

5.2.2 Communication Based on Industrial Ethernet

Ethernet is a wired data network originally designed for local data communications, primarily within office buildings. It specifies cable types, connectors, the entire physical layer, and the protocol for data transport.

Ethernet communication was developed to transmit data packets between different participants over a transmission medium. Data is transmitted in packets without a fixed access network, but a minimum bandwidth cannot be guaranteed. This means that classic Ethernet communication is unsuitable for real-time applications since insufficient bandwidth can result in data packets not reaching Ethernet components in time. For example, this effect can be seen in video streaming, where it manifests itself in reduced image resolution or audio dropouts. However, unlike fieldbus systems, Ethernet communication allows for reconfiguration during operation.

Initially, Ethernet was less suitable for industrial use due to the fragility of connectors and non-real-time bus access procedures and protocols. The so-called Industrial Ethernet standards have significantly improved these aspects, making them applicable to industrial applications, and are now gaining widespread acceptance.

In real-time industrial applications, segmenting the communication channel into a real-time and a non-real-time channel is common practice. The real-time channel operates deterministically, processing data packets within a defined communication cycle. The non-real-time channel operates similarly to traditional Ethernet communication.

Ethernet serves as the dominant foundation for Internet Protocol (IP) families, governing real-time data communications through additional layers of functionality for mediation and transport.

However, the promise of a unified communications infrastructure has yet to be realized: standardization of the Ethernet protocol for real-time applications remains elusive as extensions to the Ethernet protocol are a prerequisite. Nevertheless, ongoing efforts to develop Ethernet adaptations, such as the Time-Sensitive Networking (TSN) protocols, use various methods to achieve real-time capability. The Time-Sensitive Networking Task Group (IEEE 802.1) is now formulating forward-looking standards relevant to automation.

Table 5.3 provides an overview of current Industrial Ethernet protocols.

5.3   Wireless Communication Technology

Wireless communication is becoming increasingly common in automation applications. It is cost-effective and easy to install and operate. This advancement allows factories to explore new organizational approaches in which each piece of equipment involved in production and the products and raw materials used are equipped with individual wireless communication devices. This comprehensive interconnection offers numerous advantages, enabling the traceable coordination of production and logistics elements, even in unforeseen circumstances. Flexible coordination of this kind is only achievable through wireless communication.

However, many questions remain: What technologies are available? What are the requirements and applications? What criteria should be applied when selecting among the various technologies on the market? Which technologies are easy and cost-effective to integrate, build and maintain?

Understanding these requirements is critical to evaluating the suitability of a radio technology for wireless communications. The physical transport layer is based on the 2.4 GHz, 5 GHz, and 868 MHz bands. Various standards exist, such as IEEE 802.11 for wireless LAN, IEEE 802.15.1 for WPAN/Bluetooth, and IEEE 802.15.4 for wireless networks. However, there is still a persistent issue of bands with similar frequencies overlapping,

Table 5.3 Ethernet in industrial use

Name	Area of application	Characteristic	Wiring	Reaction time	Extension (per segment)	No. of participants
PROFINET	Drive technology and manufacturing industry	Real-time capabilities	Line Tree Star	Medium (5–10 ms)		64
POWERLINK	Transmission of process data	Real-time capabilities	Star Tree Line Ring	Fast (200 µs)	100 m per connection	240
Ethernet/IP	Control systems, manufacturing	Limited real-time capabilities	Star Ring Tree			Approx. $2^{(16*4)}$
EtherCAT	Controls	Short cycle times	Line Tree Star Ring	Very fast (1 µs)	100 m	65,535
SERCOS III	Drive technology and manufacturing	Cross-communication	Ring Line	Fast (31 µs to 62.5 ms)		511
MODBUS TCP	Basic field communication (like MODBUS, but improved)	Easy to implement, without real-time guarantee	Tree Star Ring		Unlimited with fiber optics	2^{32}
CC-link IE	Controller-to-controller and I/O communication	Support for time-sensitive networks	Line Star Ring		100 m	120 for control and 254 for field

thus partially blocking frequencies or causing interference. There is clearly a need for further standardization in this area.

There is a wide range of wireless technologies that vary in coverage, data rate, and application. Wireless product solutions typically adhere to IEEE standards but define additional specifications and certifications as well as creating proprietary distribution channels through corporate consortia.

Table 5.4 provides an overview of wireless technologies for some typical applications of automation.

Many different features and characteristics distinguish wireless networks: encryption capabilities for secure communications (range, throughput, and infrastructure), efficiency, chip size, device integration effort, cost, security, and scalability. The requirements of application groups in this area are manifold. For example, if a machine needs wireless access to quickly interface with a service technician's device, a common approach such as

Table 5.4 Overview of wireless networks for local communication

Name	Area of application	Sector	Range	Infrastructure	Efficiency	Chip size
WLAN	Broad access	Many	100 m	Router	High	Medium
Bluetooth	Product interface	Consumer	10–100 m	Point-to-point	Low	Small
ZigBee	Devices	Consumer	100 m	Connecting node	Low	Large
Wireless HART	Sensors and actuators	Process industry	250 m	Connecting node	High	Large
Industrial WLAN	Sensors and actuators	Manufacturing industry	100 m	Connecting node	High	Large
EnOcean	Smart home	Building automation	30 m	Connecting node	Very low	Large
5G/LTE	Mobile communications	Telecom communication	Large-area	Transmission tower	High	Medium

WLAN, Bluetooth or ZigBee will suffice. However, devices with specific security requirements such as a remote control for an implanted pacemaker require other methods that meet stringent security standards and work only in close proximity or in the near field.

In order to respond quickly to market demands, special interest groups have formed to develop specific product solutions. Both Bluetooth and the ZigBee Alliance leverage a global partner network of companies, universities and government agencies to drive ZigBee into the marketplace. Conversely, the HART Communication Foundation and products such as industrial WLAN are tailored more to the automation of industrial production systems, and they address specific industrial applications.

A critical success factor for wireless communication is energy efficiency, especially for wireless components that are not networked and run on batteries. Bluetooth and ZigBee offer several low-power modes when communication is not required. The EnOcean Alliance provides low-power solutions with energy harvesting functions enabling EnOcean-compatible switches to be powered by batteries or self-generated energy as the transmitters consume very little power.

Mobile communication standards such as LTE and 5G offer a wide range of opportunities for wireless communication in all sectors and applications of automation and industry. In particular, 5G offers opportunities in the 2 GHz and 3.4 to 3.7 GHz frequency ranges and allows the integration of individual specialized networks that operate in a shared physical infrastructure but are adaptable to different requirements.

Three configurations characterize networks tailored for 5G:

– Ultra-fast mobile broadband communications capable of data rates of up to 10 GB/s,
– Reliable communication for real-time applications, enabling short response times in the millisecond range,
– Data communication between multiple subscribers, facilitating connections between components in the Internet of Things.

Thus, depending on the requirements, data volumes can be small and connect many participants or allow large bandwidths while connecting fewer participants. Accordingly, transmission speed and reliability requirements can be adjusted without sacrificing other characteristics. The architecture of the 5G network is highly responsive to user needs.

5G technology is a milestone in the standardization of mobile services and an innovation over previous technologies.

5.4   Communication with Objects

The ability to communicate with objects, for example to determine their presence or identify their properties, is a capability that is often required.

Figure 5.3 schematically illustrates how objects can be wirelessly connected to a communication network via a reader. Information about the objects is stored in so-called tags and then processed within a connected system. This establishes relationships between objects, leading to applications that can for example track an object's path or continuously monitor its state. This allows objects to be uniquely identified and permits them to share and store information.

If the information coming from an object's tags is available in a communication network, it can be processed and coordinated in control units. Specific description languages are used to represent properties and characteristics in a consistent way. Since the 1970s, various methods have been developed for machine-readable labels, making it easy and inexpensive to apply identification markings, for example by printing. Barcodes or QR codes are simple and robust methods that are well established in many everyday applications. For example, barcodes can store a sequence of 13 digits, while Data Matrix codes can represent up to 1556 bytes and QR codes up to 2956 bytes. However, these codes are inherently readable by anyone and represent immutable information.

In contrast, smart labels or smart tags can identify and locate objects contactlessly through RFID (Radio Frequency Identification). RFID offers read and, in some cases, write capabilities and is available in a variety of low-cost formats. Reading ranges can vary significantly depending on the technology and the application and can range from a few centimeters to several meters.

Table 5.5 outlines the frequency ranges, their reach, and their applications.

Different versions of RFID are distinguished by the frequency used, the range, and the type of power supply. When operating in what is known as the near field, where the dis-

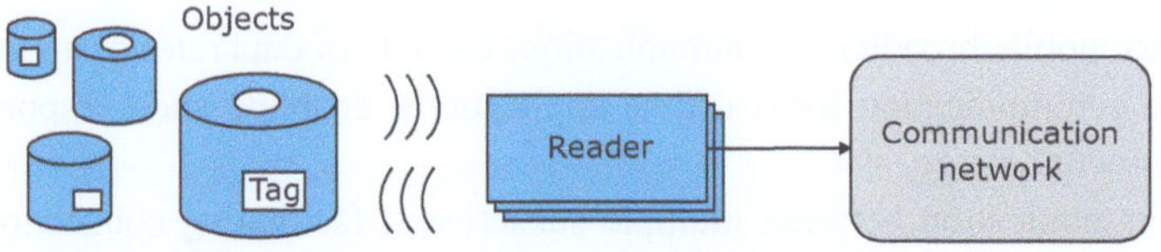

Fig. 5.3 Structure of information processing for object identification

Table 5.5 Wireless methods for object communication (overview based on DIN SPEC 91406, 2019)

	Use case	Range	Frequencies
RFID	Suited for material with a high water content	Range of a few cm	9 kHz to 135 kHz
	Use in access control	Range of a few cm	6.78 MHz, 13.56 MHz, 27.125 MHz, 40.68 MHz
	Warehousing and logistics applications	Range of several meters	433.920 MHz, 868 MHz, 915 MHz, 2.45 GHz
	Vehicle identification	Range of about 10 m	5.8 GHz, 24.125 GHz
NFC	Payment systems	Range of a few cm	13.56 MHz

tance is lower than the wavelength ($< \lambda$), frequencies below 100 MHz can result in short reaches in the centimeter range due to so-called inductive coupling. For longer distances, the far field is used, where the distance to the tag is $>\lambda$. Frequency bands in the UHF or microwave range are typically used here, allowing distances of 3 to 4 m, and sometimes more than 10 m. Coupling is based on the dipole principle using electromagnetic waves.

Different applications can be implemented according to the range and reliability of object detection. Near Field Communication (NFC) enables payment transactions at the cash register, ticket reading, and time recording within the near field communication range. At longer ranges, RFID is used for electronic product identification or location within warehouses.

In addition to the range, the power supply is an important aspect of the application spectrum. There are two different methods:

- Passive transponders are charged by induction and use this energy, briefly stored in a capacitor, for transmission.
- Active transponders require a power source in order to collect and process data.

Depending on which variants are used, very different applications are possible. Today, RFID has many practical applications. Many products incorporate RFID as an anti-theft device to detect whether a product has been paid for. Other products such as high-value pharmaceuticals use RFID to prevent counterfeiting. Even road tolls are collected through RFID embedded in vehicles. RFID can also be embedded in warehouse floors to serve as reference markers for mobile robotic systems. Because of the low cost of tags and readers, the potential applications for RFID are expected to grow in the future. The same trend applies to NFC technology, which is already widely used in payment systems.

5.5   The Internet of Things—Networking on a Large Scale

Building on the concept of the Internet, there is an increasingly robust interconnection between systems, subsystems, and individual objects. These "things" are interconnected using information technology to enable new applications.

At its most basic level, this concept is compelling because it enables cyber-physical automation systems for which connectivity between subsystems is a fundamental requirement. Upon closer inspection, however, the Internet of Things vision is complex and much more difficult to implement than its name suggests; simply having an interface to the Internet is not enough.

In recent years, numerous initiatives and significant investments have been made to establish reference architectures and system platforms. The central approach is to develop a unified architectural concept, standardized protocols, and IT components to overcome fragmented software implementations specific to certain systems and use cases.

The need for reference architectures in the industry is underscored by the increasing number of initiatives aimed at standardizing IT architectures and protocols. These initiatives aim to enable interoperability, simplify development, and facilitate implementation. However, IoT based on reference architectures is not universally applicable today. Nevertheless, there are many specific use cases in which a variety of subsystems can be interconnected. Examples of purpose-built software systems for cases of this kind include package tracking, fleet management for logistics providers, parking guidance systems in large cities, remote maintenance for wind turbines, and so on.

5.5.1   The Industrial Internet Reference Architecture

Reference architectures can guide implementation, whereby not all aspects and details actually need to be implemented. An important initiative in this regard is the Industrial Internet Reference Architecture (IIRA) from the Industrial Internet Consortium (see Fig. 5.4). Established by companies in the information and communications industry, the IIRA outlines a practical reference architecture for IoT application areas. In the meantime, a lot of research effort has been dedicated to detailed architectures, functional and information viewpoint models, and system requirement analyses in application areas.

There are no unified architecture standards based on the IIRA at present. However, the guideline helps to delineate different system domains. Many market participants offering large platforms are now aligning their structures and their terminology with the IIRA.

The IIRA integrates fundamental network concepts on the macro scale with those in the immediate local environment. The architecture introduces a proximity network that aggregates real-time data from the immediate environment and transmits it through the edge to the platform and to enterprise layers. Here, control tasks, data storage, and analysis take place in the cloud regardless of location.

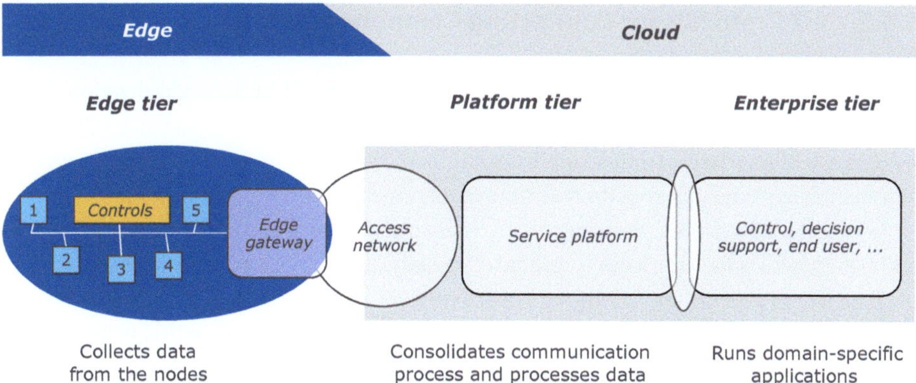

Fig. 5.4 Overview of the Industrial Internet Reference Architecture (IIRA), adapted from Industrial Internet Consortium (2019)

The platform layer includes communication processes and data processing whereas the enterprise layer manages domain-specific applications with a strong application context. In defining these tiers, the authors obviously had the tasks to be performed in mind: facilitating the transmission and collection of data and its analysis with regard to the enterprise and its domains.

The division into edge and cloud, with the possible addition of fog, stems from a metaphorical perspective on communication or IT architecture. An edge server connects devices in the field that do not necessarily need to transmit all information to the cloud but can process it within the edge and fog environments. For example, if a user interface to a machine needs to be connected via a 5G-enabled tablet, it ideally needs to be connected via an edge server responsible for that particular segment. For this reason, this connectivity is logically implemented in the same segment or cell.

The real-time capability for cloud access is somewhat limited at present, only accommodating a smaller number of processes with low time requirements. However, communication via the cloud is fast enough to react within a few seconds, even when multiple users access the service at the same time.

To allow the further processing of data or connect different edge areas, an edge gateway and an access network connects them to the cloud. Special IoT communication protocols can be used for this purpose. Finally, the data converges in the cloud. The distinction between the platform and enterprise tiers addresses aspects of data sovereignty and potentially of physical storage and processing locations that are required from an organizational and legal perspective.

5.5.2   IoT Protocols for Connecting Components

New protocols emerging from the IoT (Internet of Things) and standardization bodies specifically support technologies aligned with the IoT stack. This is because IoT spans a wide range of use cases, from a single constrained device to highly networked, large-scale automation systems that require real-time communication.

In distributed systems, ensuring cohesive interaction between automation components and the processes they perform is critical. Typically, a connectivity layer is used to distribute data and to process and coordinate distributed processes and application software. This makes it irrelevant whether a participant executes a capability itself or this is provided by a resource within the network.

The many existing and interacting communication protocols become relevant when we try to connect systems and components. Communication within an IoT stack occurs across layers using various Internet protocols that organize communication either cyclically or in an event-driven way. The implementation concepts vary, and depending on the requirements, they support aspects such as particularly fast or reliable and secure exchange of messages.

The various elements of the protocol stack designed for IoT applications are shown in Fig. 5.5. This overview categorizes protocols and standards at different levels of Internet connectivity, transport, and application.

Protocols for Internet connectivity and transport give priority to the collection and mediation of data between multiple partners. For example, the IPv6 protocol has solved the addressing problem caused by the limited number of addresses in its predecessor—IPv4—and now allows for a large IP address space. However, only a small portion of the data within a message can be used for communication as the rest is allocated to message overhead. For this reason, IPv6 is limited in its suitability for power-constrained applica-

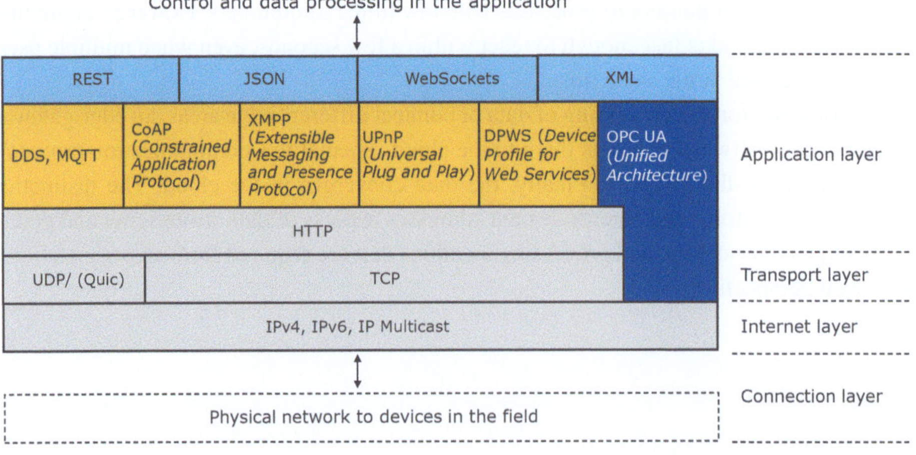

Fig. 5.5 Protocol stack of the IoT

tions such as the continuous operation of battery-powered systems. In specific scenarios of this kind, energy-efficient protocols designed for wireless communication are used.

On the transport layer, several protocols are available today. The high-performance User Datagram Protocol (UDP) is suitable for transmitting large amounts of data but lacks features to ensure transmission reliability. The newer Quic (Quick UDP Internet Connections) extends UDP and is also suitable for fast data transfer with improved reliability.

The Transmission Control Protocol (TCP) is widely used but presents challenges in hard real-time applications due to its inability to guarantee real-time capabilities.

On the higher application layers, we find service-oriented protocols that take semantics into account. This means that parts of the protocols are already assigned with a meaning relevant to an application domain, including its connections and data structuring.

These software architectures connect devices in the field by encapsulating communication between components of a distributed system and abstracting their interfaces. Different communication protocols have emerged to meet different requirements:

- DDS (Data Distribution Service): Managed by the Object Management Group, this provides reliable and powerful real-time communication between machines. It works according to the publish/subscribe principle, where a shared bus is available to publishers and subscribers, guaranteeing the delivery of messages according to various quality measures (data availability, delivery, data freshness etc.).
- MQTT (Message Queuing Telemetry Transport): This flexible TCP/IP-based messaging protocol is suitable for IoT devices and low-bandwidth networks with high latency and low reliability requirements. It provides three levels of Quality of Service (QoS), ensuring that messages are delivered once, at least once, or exactly once with additional handshakes. It is standardized by the Organization for the Advancement of Structured Information Standards (OASIS).
- Constrained Application Protocol (CoAP): Another energy-efficient protocol for resource-constrained devices and networks designed for interactive communication over the Internet. It translates HTTP for sensors and switches, simplifying the integration of machines into the IoT with low overhead. Messages are exchanged over UDP in a request-response architecture supporting two levels of quality with an optional message acknowledgement procedure. The protocol is specified as RFC 7252 by the Internet Engineering Task Force (IETF).
- XMPP (Extensible Messaging and Presence Protocol): This is an open communications protocol for exchanging text messages on the basis of TCP. It is based on XML and enables the exchange of structured data between distributed network entities almost in real time using the publish/subscribe principle. It is specified as RFC 6120 by the Internet Engineering Task Force (IETF).
- Other protocols: There are several other protocols, such as UPnP (Universal Plug and Play), which is designed to enable the manufacturer-independent control of home auto-

mation devices, routers, or printers. DPWS (Device Profile for Web Services) enables web services on automation components, i.e. hardware with limited resources.

- OPC-UA (Unified Architecture) is an established protocol in service-oriented architecture (SOA) and is an Open Platform Communications (OPC) specification. It provides functions and applications at different levels of communication on the basis of TCP/IP on the lowest level. This enables standardized, manufacturer-independent access to machines, devices, and other systems in industrial environments.
- All of these communication protocols enable new ways of networking devices, machines, and products in different application areas, but their implementation requires relatively complex software architectures.

There are several ways to connect to other systems or people in the top layer: Using formats such as JSON and XML (or the compressed version EXI), it is possible to create human-readable message protocols at the top level. In addition, software architectures such as gRPC and REST allow various services to be invoked and executed remotely.

However, these architectures are somewhat complex to implement as a software system. All components of an automation system need to be connected in order to mediate between them. This requires middleware that is centrally located and has a connection to all components.

Central data structures are necessary for locating the individual components. Even if highly decentralized information processing is intended later, a registry similar to a phone book is needed to locate components.

In any case, when information processing occurs at different levels or when approaches are distributed across multiple components, the result is the virtualization of processing. This can lead to cybersecurity challenges as this middleware establishes unified communications that could be exploited to compromise an automation system.

In general, the use of middleware leads to additional resource consumption within the automation system. The way in which messages are communicated is critical. Continuous cyclic communication with complex protocols results in high communication volumes that consume significant energy and bandwidth. The requirements of mobile applications call for solutions that consume minimal bandwidth during communication and do not require extensive computing resources and memory for middleware execution.

5.5.3   Perspectives on the Architectures and Protocols

Following Weyrich et al. (2014), architectures and protocols aim to meet different requirements, which can be summarized as follows:

- Flexibility in communication is crucial, allowing both one-to-one connections (unicast) and communication with multiple partners (multicast) simultaneously.

- Comprehensive data collection and analysis capabilities are essential for extracting information and knowledge that can be delivered as a service.
- Device management solutions should accommodate the addition of devices or changes in device configurations.
- Scalability across multiple subscribers is required in order to accommodate varying system sizes.
- IT security is imperative in all IoT domains in order to ensure privacy and foster trust in products.
- Centralized or decentralized approaches to data storage and processing play a critical role, especially in ensuring the privacy of sensitive data.

Throughout this chapter, the diverse nature of communication systems (IoT protocols and control software) has been highlighted against the backdrop of different technical systems. This has led to the emergence of a vast array of software-based communication technologies that can be combined in many different ways.

Different perspectives on architectures and their protocols are presented in Table 5.6. These perspectives, similar to those discussed in Sect. 3.2.5 regarding the virtualization of hierarchies, present three views of communication and data processing within the IoT protocol stack.

The first perspective revolves around "semantic orientation", which involves the interpretation of data and information with the goal of generating knowledge. Semantics concerns the meaning of the data within the operations, which requires an information model to interpret it. Protocols at this level must therefore establish a link to the application on the basis of an appropriate information model. OPC-UA is a milestone in that it not only transports machine data but also structures it in a technically understandable way by means of service protocols. Industries such as mechanical engineering have been involved in information modeling efforts to make OPC-UA structures semantically interpretable. This architecture includes information models that address typical industrial issues, enabling the OPC-UA reference architecture to define services or information exchange by means of industry-specific standardization (companion specifications). These industry-specific information models promote interoperability at the semantic level, facilitating a consistent

Table 5.6 Three different perspectives on the IoT protocol stack

Perspective	Aspects of the architecture	Protocols used
Semantic perspective	Mapping of specific capabilities during operation over the lifecycle	Middleware protocols such as OPC-UA
Things perspective	Internal focus on components such as sensors and actuators	HTTP and other "low-level" communication protocols such as ZigBee, Bluetooth, I/O link etc. are used for bit transmission
Internet perspective	Mapping of communication using internet protocols or protocol extensions	IP, UDP/TCP or MQTT, DDS, CoAP, XMPP etc.

function call or capability exchange. While several companion specifications exist, there is still discussion about standardizing information models and their contextual meanings at the system and component level.

The second perspective is the "things orientation", which focuses on assets such as sensors, actuators, and communication. This traditional approach within the automation industry attempts to define a bottom-up reference for physical objects and their individual data. It assigns meaning to specific data without specifying it within a comprehensive information model. Instead, it associates specific bit and byte sequences with functions, thus establishing fixed meanings within the automation system. While efficient to implement, concepts of this kind are very specific to their applications. An IoT reference architecture can provide management mechanisms and help to describe the overall communication structure. This improves the understanding of bit transmission protocols and creates illustrative models that incorporate existing communication standards to show how system components, people, and technical processes interact and process data.

The third perspective, "Internet orientation", emphasizes the use of IoT protocols for service support in conjunction with data management in the cloud and on servers. This involves the use of various layers implemented with IPv6, organizing transport either through TCP (Transmission Control Protocol), UDP (User Datagram Protocol), which allows the fast and easy transmission of simple data packets, or the QUIC (Quick UDP Internet Connections) protocol, which meets additional security requirements. In addition, advanced protocol variants such as DDS or MQTT which are specifically designed for the transmission of sensor telemetry data have emerged.

5.6   Prompts for Reflection

This chapter describes a variety of communication protocols for a wide range of industrial requirements. Many of the fieldbus systems presented for local communication are legacy systems, having been developed in the 1980s. Although they may seem outdated, they are still in use today. They are simple, inexpensive, and established in numerous products, so they are a tried-and-tested means of implementing industrial communication. With this in mind, it is understandable why the much newer methods of Ethernet or TSN (Time-Sensitive Networking) have not yet taken off: while they are gaining traction and expanding in terms of application areas, there are numerous variations, and this once again leads to a multitude of protocol standards.

The lack of interoperability between communication systems results in manual efforts to connect one fieldbus to another and adapt the software accordingly when integrating different systems.

The field of wireless communication is also expanding, although it hasn't yet reached the level of widespread adoption enjoyed by wired communication systems. The greatest hope for wireless communication lies in the 5G/6G standards, and in particular in the enablement of campus networks, allowing large companies to fully equip their factories. 5G/6G is also expected to bring many benefits to product networking.

For large-scale networking, the current approach is to rely on the existing Internet standards in order to achieve a unified connection between software systems that use these protocol stacks. However, various technical limitations limit this approach in the case of applications with moderate real-time processing requirements.

Consider the following questions and discuss the range of possible future developments.

a. Evolution of local communication systems

Why is such a wide range of different communication systems available for local communications, and why is it so difficult for new and improved communication standards to gain immediate acceptance?

What are your conclusions and expectations regarding the use of wired communication systems in the near future?

b. Fundamental limits of wireless communication systems

What are the fundamental limitations of wireless communication systems?

What are the difficulties and disadvantages of wireless communication systems compared to wired systems in industrial applications? Consider the arguments against wireless communication systems.

On the basis of what arguments could 5G/6G bring about a change and the increased use of wireless communication?

c. Effort required to implement IoT protocol stacks

What effort is required in order to implement Internet of Things (IoT) protocol stacks?

Discuss the options available for implementation as a result of the different variants of protocol stacks. What is your assessment of the potential development effort required for realizing large-scale IoT systems for practical applications involving many participants?

Further Reading

Bicaku, A.; Maksuti, S.; Palkovits-Rauter, S.; Tauber, M.; Matischek, R.; Schmittner, Ch.; Mantas, G.; Thron, M.; Delsing, J.: **Towards trustworthy end-to-end communication in Industry 4.0.** Conference IEEE INDIN, 2017. https://doi.org/10.1109/INDIN.2017.8104889

Greengard, S.: **The Internet of Things.** MIT Press, 2015. https://doi.org/10.7551/mitpress/10277.001.0001

Matheus, K.; Königseder, Th.: **Automotive Ethernet.** Cambridge University Press. 2021. https://doi.org/10.1017/9781316869543

Weyrich, M.; Ebert, C.: **Reference architectures for the Internet of Things.** IEEE Software 33, 112–116, 2015. https://doi.org/10.1109/MS.2016.20

Wollschlaeger, M; Sauter, T.; Jasperneite, J.: **The future of industrial communication: Automation networks in the era of the internet of things and Industry 4.0.** IEEE industrial electronics magazine, 2017. https://doi.org/10.1109/MIE.2017.2649104

References

DIN EN IEC 61158-1: **Industrial communication networks - Fieldbus specifications - Part 1: Overview and guidance for the IEC 61158 and IEC 61784 series.** 2020. https://doi.org/10.31030/3127706

DIN SPEC 91406: **Automatische Identifikation von physischen Objekten und Informationen zum physischen Objekt in IT-Systemen, insbesondere IoT-Systemen**, Beuth-Verlag, 2019. https://doi.org/10.31030/3114151

HMS Industrial Networks: **Marktanteile industrieller Netzwerke 2022.** Studie von HMS-Networks. Meldung in: Der Maschinenbau, 2022

The Industrial Internet Consortium: **The Industrial Internet Reference Architecture.** Version 1.9 June 19, 2019. https://www.iiconsortium.org/stay-informed-IIRA

VDI: **Industrie 4.0—Begriffe / Terms.** Status report of the working Group „Terms" des VDI/VDE-GMA FA 7.21, Düsseldorf, 2019

Weyrich, M.; Schmidt, J. P.; Ebert, C.: **Machine-to-machine communication.** IEEE Software 31 (4), 19–23, 2014. https://doi.org/10.1109/MS.2014.87

Case Study: Cognitive Sensors for Autonomous Guided Vehicles (AGVs)

6

Abstract

Cognitive sensing is regarded as a key technology for autonomous guided vehicles. However, it remains an open question on which basis and in which combination future sensor systems can be designed in order to achieve cognitive recognition and perception.

This case study will explore the variety of technologies available in the field of sensing and signal processing that can be used in autonomous guided vehicles such as those used in logistics as transportation systems or mobile robots. The focus on autonomous systems provides the framework for the development of cognitive sensing, which involves recognizing the environment by preprocessing sensor data.

This case study begins by reviewing and comparing relevant sensor methodologies. It then considers electrical hardware approaches to processing sensor data in order to implement algorithms for recognizing and perceiving the environment.

The following questions are addressed:

- Which sensor methodologies are relevant for autonomous mobile robots?
- How can a systematic selection of signal processing hardware be performed and what are the relevant criteria?
- How can a development plan for future products be sketched out despite the remaining uncertainties?

This case study provides the competence necessary for finding solutions amidst technological uncertainties. Firstly, it analyzes and evaluates the current state of the art and outlines different alternative solutions, and secondly, it presents the results in the form of a roadmap and a phased implementation plan.

M. Weyrich, *Industrial Automation and Information Technology*,
https://doi.org/10.1007/978-3-662-69243-1_6

6.1   Why Cognitive Sensors?

This case study investigates a fusion of sensor technology and electrical signal processing hardware with the aim of discussing a future generation of cognitive sensors. These sensors are envisioned as innovative products suitable for implementation in autonomous guided vehicles (AGV), for example for use in transportation.

6.1.1   What Are Cognitive Sensor Systems?

Sensors serve as the sensory organs of any technical system. However, the understanding of what constitutes a sensor and what its exact functions should be has evolved over time.

In the past, it was considered sufficient for sensors to provide measurements that could then be processed. For example, a temperature sensor might provide a voltage level that is proportional to temperature, or it could digitize that value so that a fieldbus message would be sent.

Today, cognitive sensors are expected to perceive their environment, meaning they must not only capture sensory data but also interpret it.

For example, if objects are to be recognized, they must be technically detected, and then information about these objects can be provided based on pattern recognition. Cognitive sensors must not only capture signals, but also process them in real time to enable object perception, which requires significant computing power for pattern recognition algorithms. The result is high-quality information which for example indicates the presence of a person in the observed area.

In developing cognitive sensors, the question is what sensing methods are available and how they can be used to detect objects and the environment. When a sensor is upgraded to a cognitive sensor, the perceived paths and obstacles can be communicated to an autonomous system, e.g. for path planning.

The challenge, however, is to integrate sensors with computer hardware for information processing to meet high expectations while keeping costs and technical requirements within limits.

According to industry experts such as Heumer (2021), autonomous vehicles and other products such as fully automated transport robots as well as construction or agricultural machinery are only possible through a combination of different sensing methods.

6.1.2   Structure of a Cognitive Sensor

Figure 6.1 shows a mobile transport robot that detects a person in its path and brakes or avoids the person. This requires a comprehensive perception of the environment, which is interpreted by a cognitive sensor system.

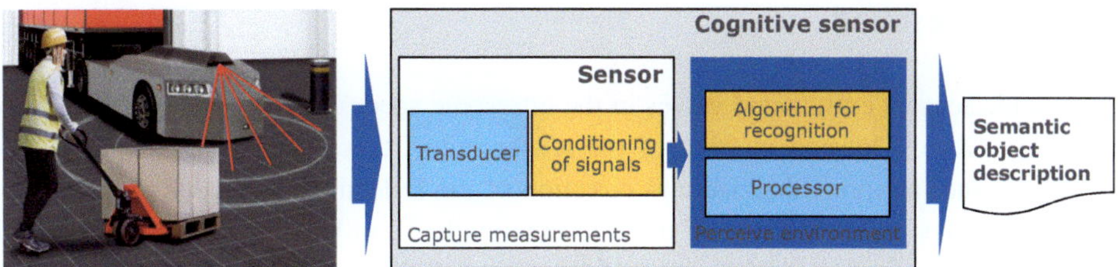

Fig. 6.1 Application of a cognitive sensor and representation of its functional groups

The cognitive sensor is a system that gathers data from the environment. The generated signals, often analog, typically need to be conditioned for further processing, e.g. adjusted to a certain voltage level. These signals are then digitized using an analog-to-digital converter to make them usable by microprocessors. Electronic hardware such as microprocessors allow compensation for non-linear characteristics, averaging, etc. Such digital signal processing often combines different measurement techniques to achieve the best possible result. The processed measurement data can then be interpreted.

The interpretation of these measurements is performed by algorithms ranging from simple decision trees to complex computational rules and AI algorithms. Depending on the captured measurements and the application, appropriate algorithms need to be chosen. For example, pre-trained neural networks are suitable for clustering image data for object recognition and providing the extracted information via a communication interface.

The output of the cognitive sensor system is an interpreted map of the environment with a semantic description of the objects.

Obviously, many of the technological aspects are difficult to assess at first. They need to be explored in detail to form meaningful combinations that can create a new product. The cognitive sensor prototypes are then designed to confirm feasibility, and customer interactions provide insights into the value proposition.

6.2 Selecting the Sensor Methods

The first step is to assess the suitability of the sensing methods. This requires an answer to be given to the following question: What are the sensing technologies and how can their application be evaluated with regard to the intended use, in particular in autonomous guided vehicles in industrial environments?

Various methodologies using different technical concepts such as ultrasound, LiDAR, radar, and cameras are widespread in sensor technology and are commercially available, as can be observed in companies such as Mercedes (2018) or Continental (2023). These sensor methodologies are discussed and compared in the following.

6.2.1   Ultrasound

Ultrasonic sensors emit sound waves in the ultrasonic spectrum (> 30 kHz, often around 55 kHz). Objects located in front of the sensor reflect these sound waves. The sensor measures and interprets the size and direction of the reflection. Typically, an ultrasonic sensor operates within a range of three to maximally 10 m but provides minimal depth information. Depending on the arrangement, size, and direction of the sound waves, this limitation leads to inaccurate lateral information and a narrow field of view, However, ultrasonic sensors are relatively weather-resilient, with only rain and snow causing measurements to be inaccurate.

Typically, cycle times of 60 milliseconds are achieved, meaning that the sensor transmits its measurements through its interface every 60 milliseconds, resulting in data rates of approximately 0.1 Mbps. The cost of ultrasonic sensors ranges from a few cents to several euros.

6.2.2   Radar

Radar—short for Radio Detection and Ranging—is a method of radio-based detection and range measurement. It sends out high-frequency electromagnetic waves and measures the reflections generated by the environment. Radar sensors are used to detect both static and moving objects.

The following characteristics apply to automotive radar systems: Authorized frequencies typically range from 77 to 81 GHz, providing high depth accuracy. However, radar forms a "beam," i.e., it fans out, resulting in limited lateral accuracy. Depending on the field of view, short-, medium-, and long-range radar can be achieved, whereby inaccuracies can occur with small or nearby objects. Landmark detection can be largely independent of weather and light conditions. Data rates typically range from 0.1 to 15 Mbps.

The principle of a radar system is illustrated in Fig. 6.2, with cycle times averaging about 66 milliseconds. Radar systems used in vehicles, for example, have a range of 0.2 to 250 m. Costs vary and can reach several hundred euros depending on the specifications.

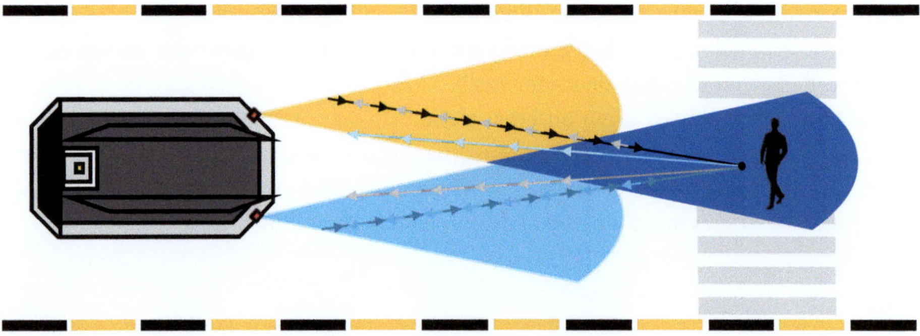

Fig. 6.2 Radar system in operation with a mobile platform

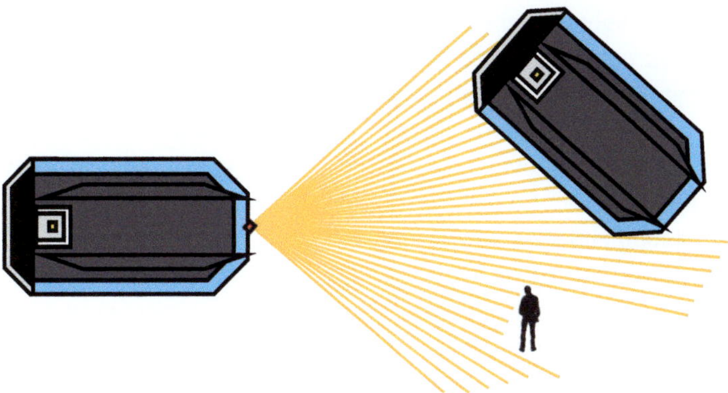

Fig. 6.3 LiDAR system for object detection

6.2.3 LiDAR

A LiDAR (Light Detection and Ranging) sensor emits laser light and measures its reflection from objects in the environment (see Fig. 6.3).

Typical ranges for LiDAR sensors used in vehicles are approximately 30 to 200 meters, and they operate in the non-visible spectrum at a light frequency of 900 nm. LiDAR achieves very high depth and lateral accuracy due to the precise alignment of the laser beam.

LiDAR uses rotating mirrors or arrays of micro-mirrors arranged in fields (MEMS or Micro Electro Mechanical Systems) to provide fields of view ranging from 40 to 360 degrees. This makes a panoramic view possible. Scanning rates vary significantly depending on the design concept and currently range from 10 to 20 frames per second, resulting in sampling rates of about 10 kHz per point in MEMS-based systems.

LiDAR performance is affected by weather conditions: Rain has a moderate effect depending on its intensity, while fog can reflect the laser beam and distort the detection. The design of the laser for LiDAR systems has to observe stringent requirements in terms of eye safety.

Affordable LiDAR systems in the double-digit euro range for vehicles and mobile robots are increasingly being developed and marketed. However, high-precision devices with exceptional performance are still relatively expensive.

Combining LiDAR with other sensor systems has the potential to leverage advantages and reduce costs.

6.2.4 Computer Vision

Computer vision is a field that involves teaching machines able to understand and interpret visual data, allowing them to perform advanced tasks and make decisions based on the information extracted from images or video streams coming from cameras.

Cameras receive light from their surrounding environment. They can achieve very high resolutions, detect colors, and operate in low-light conditions. Depth accuracy can be increased by combining two or more cameras into a stereo or multi-camera system.

Camera systems offer detection ranges from a few meters to several hundred meters and have the potential to provide an angle of view of up to 180 degrees which can be adjusted by means of optics. Cameras have different sampling or evaluation cycles depending on the resolution and the sampling frequency of the system, but they can generate a significant amount of data (in the order of 250 Mbps).

Cameras are susceptible to contamination. The cost of camera systems varies widely depending on resolution, sensitivity, and number of cameras and starts from a few euros.

6.2.5   Which Sensor Should Be Utilized?

The above discussion has demonstrated the availability of several promising sensing methods that can be used individually or in combination. A systematic evaluation and assessment must be performed in order to select the right sensor as the technical requirements of the field of application play a crucial role.

For transportation systems, it is apparent that the sensors should possess a range appropriate for the application and ideally provide information on depth and lateral accuracy. Their independence of the weather significantly enhances the applicability of autonomous guided vehicles, especially in outdoor storage areas, as they can even function in rain, fog, or snow.

The total system cost is critical and should be as low as possible and certainly not exceed critical thresholds as this could make a cognitive sensor unsellable. When evaluating costs, it is important to remember the rapid technological advances made in LiDAR. Significant investments are currently being made to develop automotive LiDAR, which is recognized as a key technology for automated driving. As a result, new LiDAR sensors manufactured at significantly lower cost are emerging due to the implementation of newly developed technologies. In turn, the high-volume production typical of the automotive industry leads to economies of scale that reduce unit costs.

Table 6.1 summarizes the key characteristics of sensor systems and provides a comparative analysis.

It soon becomes apparent that, due to their limited technical capabilities, ultrasonic sensors can at best act in a supporting role, for example in parking or docking procedures. Radar systems are able to detect objects in the environment but do not provide enough information for them to be used as the sole source of information. LiDAR systems are promising due to their technical characteristics, but their universal practical application is still relatively expensive.

Table 6.1 Comparison of measurement methods for sensor systems

Criteria	Ultrasound	Radar	LiDAR	Camera
Range of coverage	◔	●	●	◕
Depth information	◔	◑	●	◕
Lateral precision	◔	◑	●	◕
Weather-independence	◑	●	◑	○
Color detection	○	○	○	●
Cost efficiency	●	◑	◔ to ◑	◕

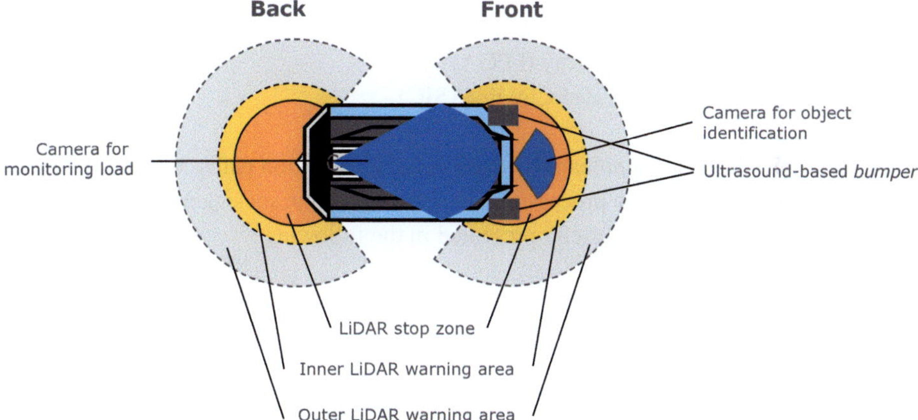

Fig. 6.4 An autonomous guided vehicle (AGV) with multi-sensor capabilities

Cameras offer excellent capabilities but are weather-dependent. Image analysis is in particular susceptible to disturbances such as contamination, which may impair reliability. However, in the industrial context of AGVs, the reliability and precision of sensor systems is paramount. Seamless data acquisition is critical in order to minimize risks to operators and other components of a facility. Also, AGV systems used in an industrial context serve a wide range of application scenarios, each of which requires the sensor system to provide different functionalities.

This makes it necessary to consider combining different sensors. A plausible combination for an AGV is shown in Fig. 6.4.

The question as to which sensor method to favor is not easy to determine as different methodologies can be used individually or in combination.

6.3 Which Electrotechnical Hardware Is Suitable for Processing Information Quickly?

An essential aspect in the conceptual design of a cognitive sensor is the electronic hardware used for information processing. This hardware serves as a pivotal point in the conceptualization as the information processing capabilities and the associated costs ultimately determine the acceptance of the overall concept.

In the following, the alternatives for the hardware of the processing units are examined and compared. According to the state of the art, the following options are available:

– Central Processing Unit (CPU)
– Graphics Processing Unit (GPU)
– Field Programmable Logic Array (FPGA)
– Application-Specific Integrated Circuit (ASIC)

Each alternative leads to significantly different implementation paths, which must in turn be adapted to the information processing requirements of the sensing method in question. These technical solutions are briefly presented in the following.

6.3.1 Central Processing Unit (CPU)

A CPU is the main component in a computer and is responsible for receiving instructions from a program and executing mathematical operations. Initially, CPUs had a single processing core. Modern CPUs consist of multiple cores that allow multiple instructions to be executed simultaneously. The number of processor cores typically ranges from single to double digits, with core clock speeds of 4.8 GHz and a power consumption ranging from 6 to 220 watts. CPUs can be programmed in both high-level language and assembly language. They are optimized for fast sequential processing and are known for their ease of programming and their versatility.

6.3.2 Graphics Processing Unit (GPU)

Initially introduced to relieve CPUs of graphics processing tasks, GPUs consist of serially connected parallel processing units that allow the efficient handling of highly parallelizable processes (see Fig. 6.5). The advances in terms of high-resolution image processing and computer vision for object identification have made GPUs more and more important. Many algorithms in various fields of application benefit from GPU parallelization, thus extending their use beyond image recognition to include machine learning and data processing. GPUs are optimized for computing large-scale vector/matrix data but have a high power consumption. The number of cores ranges from several hundred to several thousand

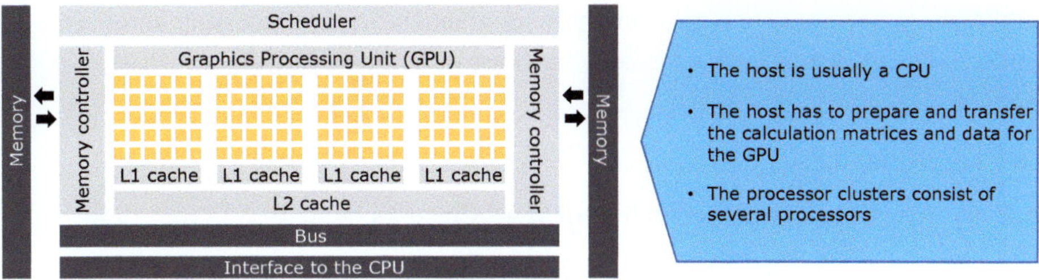

Fig. 6.5 GPU design—sketch of an array of the parts of a graphics processor

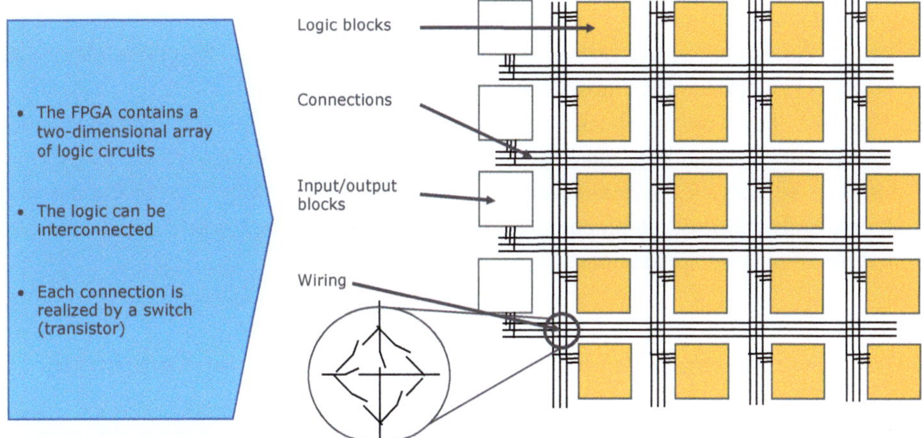

Fig. 6.6 Example showing an FPGA architecture

with typical clock speeds of between 1 and 2 GHz. Power consumption varies from low to very high depending on the size.

6.3.3   Field Programmable Gate Arrays (FPGA)

Field Programmable Gate Arrays (FPGAs) can be implemented as flexible, reprogrammable, and versatile devices containing 100,000 to 10,000,000 system gates. These devices include a matrix and/or gates and flip-flops that can be interconnected to operate in an application-specific manner. Clock speeds range from 20 to 1500 MHz, with power consumption ranging from 2.5 to 40 watts. Figure 6.6 outlines the concept of an FPGA and how it can be programmed by setting the interconnections.

FPGAs are therefore a type of "programmable hardware". Their applications include specialized analysis and real-time control requirements such as network analysis, motor control, and industrial image processing. FPGAs are well suited for signal processing algorithms because they can implement parallel digital filters. Although highly adaptable,

FPGAs are comparatively more difficult to program than other processing units, requiring in-depth knowledge of logic gates and hardware description languages such as VHDL or Verilog.

6.3.4 Application-Specific Integrated Circuits (ASIC)

ASICs (Application-Specific Integrated Circuits) are highly specialized electronic components which, in a similar way to FPGAs, combine logic elements to build complex circuits for specific applications. Once created, an ASIC cannot be modified because it is designed and manufactured specifically for its designated application. ASICs only consist of the blocks necessary for optimal operation and are therefore optimized for performance and/ or power consumption. The chips are fast and highly efficient.

However, they have drawbacks such as complex development—for interface adaptation, for example—and high manufacturing investment costs. ASICs are used in various fields such as robotics, electronic devices, and medical technology. Due to the complex and time-consuming development process and high investment costs, ASICs are best suited for large-scale applications.

6.3.5 Comparison of the Hardware Used for Information Computing

The information processing hardware must be tailored to the context, i.e. the sensor methodology and the application. Algorithms have different real-time processing requirements, and the challenges of parallelizing them differ depending on the algorithm. In turn, the application context imposes requirements on the processing unit—such as cost constraints—or on the power supply provided at the installation site.

To be able to make a selection, a comparison must first be performed on the basis of the characteristics of the hardware systems. The following criteria play a role in selection:

- The real-time capability of the algorithms implemented is a critical feature and is potentially critical if information processing is too slow.
- Programmability of the algorithms is essential in order to avoid significant development effort. Scalability for increased performance requirements in future developments is also important.
- Size, form factor, and power consumption are practical aspects which can be relevant for their application.
- The total production cost of the system is critical for a future product and is derived from the hardware costs combined with the costs of development.

Table 6.2 compares each processing unit and its properties.

Table 6.2 Comparison of the processing units

Criteria	CPU	GPU	FPGA	ASIC
Processing speed (real-time capability)	◑	●	◕	◕
Small, compact design	◕	○	●	●
Simple programming	●	◑	◔	○
Low energy consumption	◐	◕	●	●
Reasonable costs	◑	◕	◑	◔

The illustration shows a varied assessment. While CPUs are commonplace, they exhibit weaknesses in terms of the real-time capability of specific mathematical computations. GPUs are powerful but difficult and complex to program. FPGAs offer numerous advantages but require comprehensive expertise for programming and may not be suitable for all algorithms. ASICs require substantial know-how and are only cost-effective when used in large quantities.

A definitive decision favoring one approach over the other cannot be made across the board. Since algorithms cannot easily translate into one specific approach, an extensive analysis is often necessary in order to estimate performance against data flows and computations.

The alternatives should be weighed against the specific technical requirements. In addition, it may be beneficial to test several approaches at the same time for the sake of clarity. These methods can be combined in a hybrid architecture to compensate for weaknesses and achieve a more robust architecture—for instance by utilizing a CPU for general-purpose computation alongside GPUs or FPGAs to accelerate specific algorithms.

The choice of combination impacts both technology design and economic considerations. If the optimal combination of a heterogeneous hardware architecture or a combined cognitive sensor system is not immediately apparent, experimental setups and prototypes can be valuable ways to determine it.

6.3.6   Technical Requirements of AGVs

AGVs—also referred to as transport robots—are increasingly being deployed in industry to provide greater flexibility and mobility in intralogistics, which is moving away from traditional linear assembly lines. Instead of installing conveyors or similar mechanisms, these systems use autonomous vehicles that can navigate, avoid obstacles, and adapt to dynamic operational needs.

These transport systems receive a destination from a central fleet management system and then compute a path from their current location to the destination. However, since these paths are initially based solely on start and end coordinates, mobile obstacles such are other vehicles, materials, or people cannot be accounted for in advance, even if a map of the facility is available. For this reason, AGVs must continuously sense their environment, detect obstacles, and adjust their trajectories.

If the path is blocked, several solutions can be attempted. A vehicle can choose to navigate around the object while maintaining a certain minimum distance. If this is not possible, It may choose an entirely different route or request human assistance. Otherwise, it may pause operations, potentially leading to a standstill.

In the simplest scenario, ultrasonic sensors or basic LiDAR scanners can be used to detect and maintain minimum distances without contact. Accurate obstacle classification is critical to route planning. It makes a significant difference whether the path is temporarily blocked by another passing robot or is obstructed by a permanent obstacle.

This is where the application of cognitive sensors comes into play. They generate higher-order information from sensor data that directly influences the decision-making process. However, the integration of these sensors into transportation systems requires business considerations in order to decide on a concept that is of value to manufacturers. Comprehensive assessments of the needs and requirements of the field are necessary for practical implementations.

For the example of an AGV presented here, the aim is to develop a compact sensor capable of detecting objects and people in its path while providing additional insights into the nature of obstacles, thus essentially classifying them, for example by indicating "person crossing ahead".

6.4 Developing an Approach for a Product Configuration

When developing a cognitive sensing methodology for pathfinding and navigation in an AGV, a stepwise approach is recommended. This is due to the iterative nature of the specification of a new product in the presence of uncertainties with regard to technological advancement.

The development of a cognitive sensor system can be structured using a phased development plan. It is also probable that new insights may emerge during the development process, necessitating revisions to the initial plan.

6.4.1 Development Options and Selection of Technologies

In the conception and planning of developments, it is essential to make assumptions about the technological options and the prospective field of application of a future product.

The following assumptions and observations can be outlined:

– It is expected that AGVs will increasingly penetrate the manufacturing industry or even achieve widespread adoption, making this market an attractive growth sector.

- LiDAR sensors and cameras are becoming increasingly cost-effective, and this trend is expected to continue. In particular, there are indications that a breakthrough in low-cost LiDAR sensor solutions is approaching.
- The global push for the development of AI algorithms is expected to simplify and reduce the cost of using cameras in conjunction with GPU-based LiDAR signal processing.

The question as to whether the monitoring of open spaces in a factory should be carried out by permanently installed sensor systems or by sensors mounted decentrally on the transport systems is still being discussed.

Fixed sensor installations offer the advantage of comprehensive knowledge of the open pathways, enabling better utilization compared to an approach that relies on the limited knowledge of one or more autonomous systems. Conversely, autonomous units offer much greater flexibility in technology selection and are particularly advantageous for multiple vendors since individual systems can operate independently without the need for efforts to integrate them into a unified system.

How can the previously outlined considerations for the selection of a sensor method and a hardware for this application area be concretized?

First, the selection criteria must be specified based on the application and the expected requirements. Next, a weight is assigned to each criterion. These attributes are given higher or lower weight factors on the basis of their relative importance.

Table 6.3 shows the weightings and scores for the criteria identified previously along with two additional criteria which are directly relevant for the application (robustness and market acceptance).

The results of the evaluation shown here are relatively close to each other. Considering the lack of consensus among the experts, none of the sensor methods takes the lead as each has its pros and cons. However, a ranking does appear, with LiDAR at the top, closely

Table 6.3 Comparison of the sensor methods using a weighted sum

Criterion	Weight	Ultrasound	Radar	LiDAR	Camera
Range of coverage	2	1	4	4	3
Depth information	2	1	2	4	3
Lateral precision	1	1	2	4	3
Independence of the weather	2	2	4	2	0
Color detection	1	0	0	0	4
Cost efficiency	3	4	2	2	3
Robustness for the area of application	3	4	2	3	2
Expected market acceptance	2	1	1	4	3
Sums		**35**	**36**	**47**	**40**

Weight: 1 (low) to 3 (high/important); rating: 0 (poor) to 4 (good) according to the Harvey Balls classification in Table 6.1.

followed by camera systems, mainly as a result of the technological advances made in recent years.

For this scenario, LiDAR seems to be the preferred and most cost-effective system. The use of camera systems is also important when considering the potential improvement of route planning through obstacle classification. The combination of both methods in the context of vision could simplify image recognition and object classification as the complex classification of camera images is facilitated by the depth profiles of LiDAR.

Ultrasound and radar technologies are virtually equivalent but rank significantly after LiDAR, making their use less appropriate.

For the selection of the processing unit, it is assumed that LiDAR sensors will be used first and cameras will be added later. For the data processing of a LiDAR sensor, the use of a CPU seems to be feasible—assuming that the LiDAR system in use already includes signal pre-processing based on ASICs and does not require any new development. Choosing a LiDAR system without preprocessing capabilities is possible but would require significant investment in order to develop specialized LiDAR preprocessing FPGA or ASIC.

Programming a CPU is relatively cost-effective, and building the expertise is easier compared to the knowledge required for an FPGA or ASIC. In addition, a CPU can be expanded later using GPUs if complex computer vision algorithms need to be applied.

It is obvious that the choice of sensor and processing unit depends on numerous factors. While the systematic selection of sensors was based on a weighted sum in the table, this does not guarantee a correct decision due to the complexity of the technical interrelationships. Factors such as the development effort or the acceptance risk for the final product also play an important role. This judgment is strongly influenced by specific circumstances such as the competence of the development team and the intended target market.

6.4.2 Developing a Multigenerational Plan

A multigenerational plan distributes investment, development effort, and risk over a number of manageable steps. Between these steps—also known as generations—there is still room for reevaluation and readjustment. As new knowledge is gained or older knowledge is disproved, subsequent generations can be adapted. In addition, the functional interim solutions that emerge can be marketed as products, thus providing economic support for further research and development.

These interim products can later be introduced as simple, low-cost versions, thus expanding the customer base. A multigenerational plan for a cognitive sensor for AGVs is shown in Fig. 6.7. Different development generations are shown horizontally. Each generation is described in a column aligning goals, methods, hardware, functionalities, required knowledge, development tasks, and potential technology risks during the development process.

	Generation 1	Generation 2	Generation 3
Target image	Simple object detection	Object detection and classification	Combination of multiple data sources with decentralization of data collection
Sensor method	• LiDAR	• Camera	• Fusion of LiDAR and camera
Hardware	• CPU	• GPU	• Heterogeneous processing unit
Functions	• Obstacle detection • Motion detection • Evaluation using conventional algorithms	• Obstacle classification • Evaluation by AI algorithms	• Decentralized data acquisition and multi-sensor fusion
Required knowledge	• CPU programming	• Knowledge AI • GPU programming	• Knowledge AI • GPU programming
Development tasks	• Implementation of the evaluation	• Implementation of customized AI algorithms	• Communication infrastructure • Development of sensor fusion algorithms
Technological risks	Low	Moderate	Increased

Fig. 6.7 Multigenerational plan for a cognitive sensor for AGV in transportation applications

Generation 1

The starting point for the Generation 1 cognitive sensor system for navigation is a simple LiDAR sensor that determines whether the path is clear or blocked. This sensor is a purchased component that can easily be integrated into proprietary software for environmental analysis and path planning. On straight paths, this allows for early trajectory adjustment. Using CPU-based integration, the LiDAR sensor readings can be processed and interpreted to extract information about blockages and optimize the route.

Generation 2

In Generation 2, a camera and a GPU implementation can be added for information processing. Pre-trained algorithms for image recognition and object classification are available in software libraries. Only application-specific knowledge is required for the use of these libraries. The use of GPUs requires a moderate amount of knowledge allowing the development team to deploy AI-based algorithms. However, this development effort will pay off in subsequent generations that build functionality on GPU hardware.

Generation 3

Generation 3 finally realizes the cognitive sensor system, which is formed by fusing LiDAR and camera data. The processing unit hardware is enhanced to adapt to limited space, power management, computational operations, and algorithm parallelization and possibly requires ASIC development. The improved processing power enables the use of advanced AI algorithms that continuously learn from ongoing operational data in order to improve their cognition. It is conceivable that the combination of LiDAR and camera data could provide advanced insights into the environment. Feedback from pilot customers using the early generations guides and implements improvements. These customers also serve as pilot users for the new cognitive sensors.

As awareness and acceptance grows, mass production of this generation will become feasible. Mass production and lower unit prices will enable additional product approaches such as the deployment of sensors in manufacturing facilities alongside AGV. Installed cognitive sensors can detect the current state of the open driving area as well as the type and location of obstacles on the surface. This data is then made available to a central fleet management system for route planning.

6.4.3 Concept and Roadmap for Management Communication

To facilitate management's entry into the topic, a management communication plan should be formulated. The concept of the cognitive sensor in its final state should be shown first and then the way to achieve it should be explained step by step. Detailed information can be integrated into these explanations. The overall concept of the cognitive sensor after the completion of Generation 3 is shown in Fig. 6.8. It shows the main components: CPU, GPU, and purchased components such as LiDAR, ASIC, and camera. The schematic elements represent the circuit design, software development, and system integration tasks to be performed within the development process. This diagram can be used to quickly discuss the rationale behind the final solution without getting bogged down in a multitude of technical details. However, the block diagram also draws attention to the data interfaces that have not yet been discussed and for which a standard industrial solution should be used.

A development road map based on the multigenerational plan as shown in Fig. 6.9 is a reasonable way to illustrate the approach.

However, the generations outlined for management should also be supplemented with an assessment of feasibility with the existing team and a risk assessment.

In Generation 1, major risks are not anticipated, and all that is required is a manageable skill-building phase for a development team. The hypothesis that LiDAR sensors will

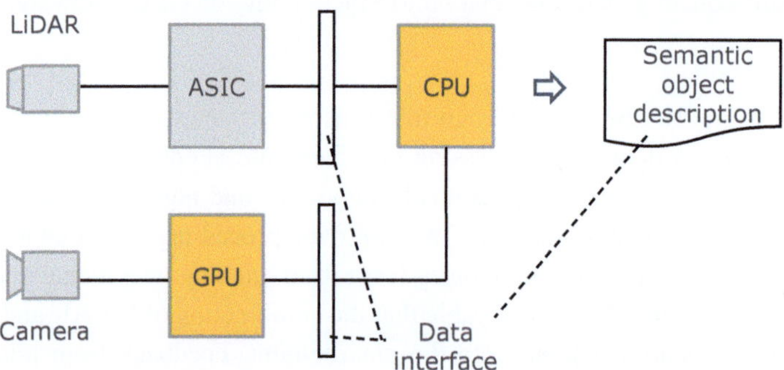

Fig. 6.8 Concept of a cognitive sensor system in the last phase of development

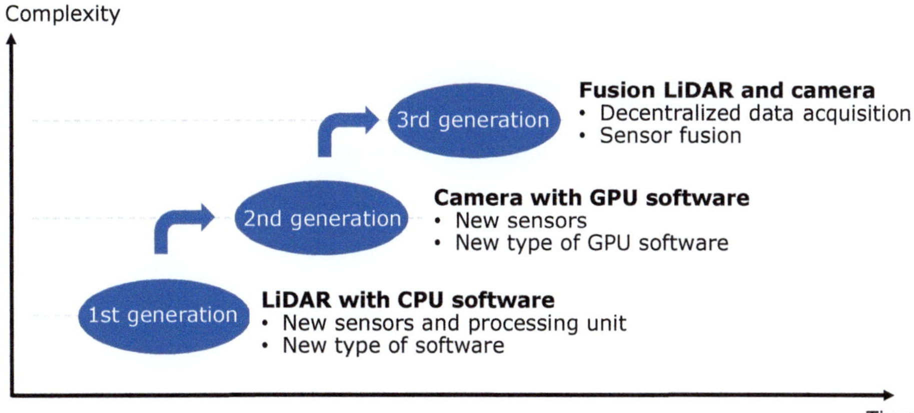

Fig. 6.9 Roadmap for development

continue to decrease in price or be used in AGV for other applications forms the basis for selecting this sensor technology.

For Generation 2, knowledge in the field of AI algorithms and GPU utilization needs to be developed. Overall, there is only a moderate risk regarding the applicability of software libraries as numerous libraries are already available.

Generation 3 poses a higher risk as additional expertise in ASIC hardware design is required. However, the development costs could potentially be funded from the profits of previous generations.

The generation plan outlined here could be challenged and expanded in further discussions. It is also beneficial to consider future generations that go beyond the considerations presented here. Possibilities include the standardization of data interfaces we mentioned earlier, the adoption of a cloud-based computing system for information processing, and the development of advanced AI methods for environmental perception.

6.5   Prompts for Reflection

This case study illustrates the complexity of decision-making in technology selection for new products and the wide range of possible alternatives that can result in cutting-edge automation products.

After careful reflection and the completion of numerous developments, identifying the pros and cons is often a straightforward matter. However, when technologies are still young and emerging, it remains uncertain which combinations will meet the technical requirements while being successful as products. It takes imagination and technical intuition to identify combined pathways. Above all, it requires systematic work to build these possibilities and evaluate them on the basis of use cases.

Please research and consider the following issues:

a. What other technologies and solutions do you envision for the realization of cognitive sensors?
b. How can these be linked and combined in different ways using the technologies outlined here?
c. Do you envision alternative directions for creating a novel sensor product for the application? How do you assess the opportunities and risks?
d. What alternative approaches can you consider for generation development and expertise building?

After researching and considering these questions, please reflect on your insights and findings.

Further Reading

Giacalone, J.-P.; Bourgeois L.; Ancora, A.: **Challenges in aggregation of heterogeneous sensors for autonomous driving systems.** *2019 IEEE Sensors Applications Symposium (SAS)*, Sophia Antipolis, France, 2019. https://doi.org/10.1109/SAS.2019.8706005

Ignatious, H.A.; El- Sayed, H.; Khan, M.: **An overview of sensors in autonomous vehicles.** Procedia Computer Science, Volume 198, 2022. https://doi.org/10.1016/j.procs.2021.12.315

Qu, J.; Barton, D.; Gönnheimer, Ph.; Pinsker, F.; Kufer, D.; Fleischer, J.: **Self-Aware LiDAR Sensors in Autonomous Systems Using a Convolutional Neural Network.** Procedia Manufacturing, Volume 52, 2020. https://doi.org/10.1016/j.promfg.2020.11.010

References

Continental: **Autonomous mobility - Camera, lidar, radar and control units provide the necessary information for highly automated driving.** Firmenschrift. https://www.continental-automotive.com/en-gl/Passenger-Cars/Autonomous-Mobility/Enablers (Abgerufen Mai 2023)

Heumer, W.: **Into the future of autonomous cars with LiDAR.** (in German). VDI Nachrichten, Issue of 28.10.2021, VDI-Verlag, Düsseldorf, 2021

Mercedes: **The sensors: Strongest in a team.** (in German) Company Magazine, Mercedes-Benz Group Media. Juli 2018

Case Study: IT Integration of Industrial Production Systems

Abstract

The IT integration of different automated production systems and their components is a significant challenge for the industry. It is often associated with high costs and numerous technical shortcomings due to the lack of interoperability of the IT and automation components. For this reason, this case study discusses a case study on the IT integration of automation in industrial production which aims to achieve interoperability between production systems.

The case study begins by exploring the requirements and then looks at possible approaches to IT architectures for the integration of these different subsystems. It covers the following topics:

- What are the specifics of IT integration in production automation?
- How can interoperability help to achieve monitoring and an optimal interaction between subsystems?
- How can service-oriented architectures be used for IT integration?
- What are the technical capabilities of the OPC UA architecture?
- What other IT architectures are there in automation, and which standards are relevant?

This case study provides some preliminary insights into the use of IT architectures in production automation and serves as a starting point for a deeper exploration of the IT integration of production systems, factories, industrial facilities or manufacturing plants.

7.1   Challenges of IT Architecture in Production Automation

The implementation of industrial automation for production systems is subject to numerous constraints and requires significant investment and engineering effort. Industrial users expect new facilities to increase competitiveness, achieve high efficiency, availability, and quality, minimize errors, and provide the flexibility to produce a wide range of products in an agile manner.

The automation of production systems is characterized by the interaction of technical subsystems from different manufacturers that must be integrated in order to perform complex processes and automation tasks.

Users of industrial automation systems aim to increase their competitive advantage and ensure high customer satisfaction. In addition to the primary requirements of the production or manufacturing process, an IT system architecture imposes additional demands on interoperability:

- Implementation of seamless interaction and smooth operation of technical subsystems from different providers.
- Easy integration of modular components, enabling their automation technology to be easily assembled, modified and operated (Plug-and-Produce).
- Realization of plant operation ensuring high quality and optimal operational processes with the help of data transparency.

Figure 7.1 gives a schematic outline of an industrial production system and includes the tasks which automation technology has to fulfil. It illustrates the wide range of activities that IT and software are to support: the large- and small-scale coordination of different equipment, numerous subsystems, and logistics on a large and small scale and the man-

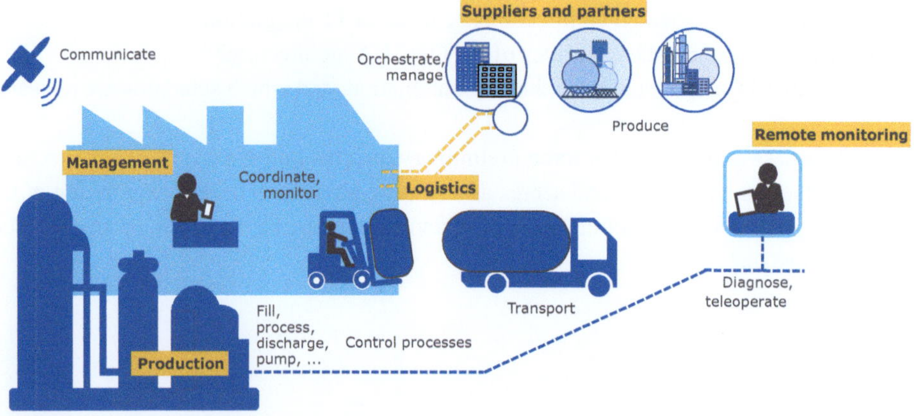

Fig. 7.1 Tasks of automation technology in industrial production

agement and diagnosis of alarms and malfunctions. To achieve this, many different systems must be able to communicate, exchange data and information, coordinate actions, etc.

In addition, industrial companies naturally want technology that provides long-term investment protection—i.e. technology that will be available for many years—as production systems often have a lifespan of more than 20 years. For automation technology, this results in a wide range of requirements for systems, components, and development tools, including

- Modularization of control and process technology through standardization of IT and its interfaces
- Establishment of system-neutral engineering through standardized functional modeling
- Provision of IT tools for commissioning, testing, and validation across manufacturer boundaries.

This is why manufacturers, industrial suppliers, and vendors of components and automation technology have been working together for years to establish standards. Cross-industry cooperation is being pursued in order to advance interoperability and the reuse of subsystems as well as the provision of exchange formats for software and data.

From the point of view of industrial automation, a distinction can be made between two trends in integration:

- The horizontal integration of subsystems to connect sensors, actuators, controllers etc. on the shop floor
- The vertical integration of production in the sense of connecting manufacturing facilities, plants, and logistics systems throughout the whole enterprise.

Many different approaches, concepts, standards, and IT solutions are available for facilitating interoperability in this area and reducing the considerable integration effort in practical terms. As the communication and information technology capabilities available for the networking of systems increase, entrenched structures can dissolve, allowing for novel implementations.

An IT architecture is selected and implemented by plant operators, i.e. companies that use automation technology in production. The goal is to integrate automation solutions on the information technology level. When all systems are compatible and data exchange takes place automatically, i.e. without the intervention of others, we speak of interoperability.

Some vertical and horizontal approaches to integration are described below to give readers an idea of how they work, followed by a SWOT analysis from a manufacturer's perspective which discusses strengths, weaknesses, chances, and potential risks.

7.2 Example: Horizontal Integration of a Manufacturing System Based on a Service-Oriented Architecture

An experimental setup from discrete manufacturing is used here to illustrate horizontal integration. The system consists of six manufacturing modules for the mechanical processing and handling of workpieces. Four conveyor belts (logistics modules) facilitate the transport of workpieces between the production modules.

An IT architecture is required which integrates different control software to ensure coordination. This allows the entire system to be controlled flexibly and simply. The virtual prototype of the production system, a schematic representation of the modules, and the electrical concept are depicted in Fig. 7.2.

The manufacturing and logistics modules, each equipped with its own microcontroller for control purposes, are interconnected by a fieldbus that provides horizontal communication. Each module is equipped with autonomous control, something which enables decentralized coordination.

The question arises as to how the electrical system components can be seamlessly integrated into existing facilities or retrofitted without extensive adaptation of the control software. A cyclic control system based on a programmable logic controller (PLC) could be used for this purpose. However, in a setup of this kind, all sequences would be programmed on the basis of a linking logic or a finite state machine, for example: "Stop the conveyor and start the drill feed when the presence sensor triggers". This type of programming is based on a contact plan or a procedural language, corresponds with the current state of the art, and would work with a centralized, cyclical control system (PLC).

The major drawback of this centralized cyclical control is that any change, such as the addition of a new manufacturing component, requires the underlying linking logic to be adapted. This results in manual reprogramming of the PLC.

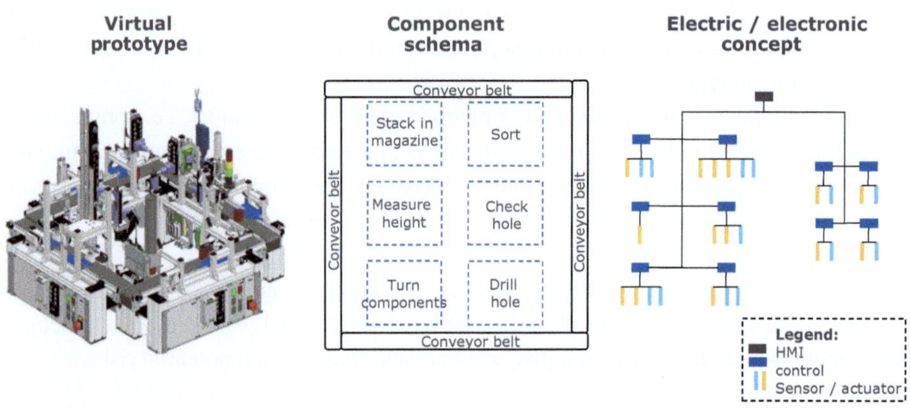

Fig. 7.2 Example of a modular production system with a component diagram and an electrical engineering concept

Another viable alternative is to leverage event-driven control paradigms implemented by means of a service-oriented architecture (SOA).

7.2.1   Service-Oriented Architecture Deployment (SOA)

A service-oriented architecture (SOA) is an IT architecture concept designed for the organization and provision of services. It facilitates the flexible utilization of distributed functionality and is considered, according to DIN SPEC 16593-1 (2018), to be one of the technological cornerstones for enabling adaptability and agility in information technology processes.

A prerequisite for SOA is that services are made available as encapsulated control software with standardized IT interfaces. These services can then be utilized on an IT platform and interconnected.

If the IT architecture of the aforementioned manufacturing facility meets this requirement, the corresponding services can be provided. This means that each of the microcontrollers associated with the production modules runs a control that communicates with other control microcontrollers over a fieldbus using a standardized protocol.

In an SOA, three roles of participants can be distinguished:

- "Service providers" offer services to service consumers.
- "Service consumers" utilize the services provided by service providers.
- "Service brokers" manage and coordinate services, whereby service providers register with them and can then be discovered and utilized by service consumers.

The interaction of these roles in an SOA is illustrated in Fig. 7.3.

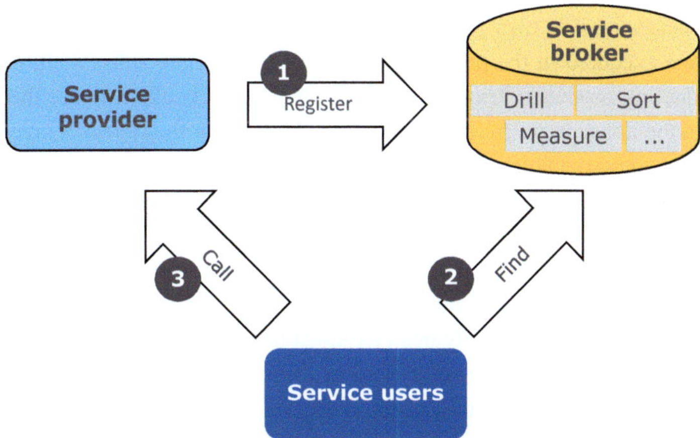

Fig. 7.3 Principles of interaction between the elements of an SOA

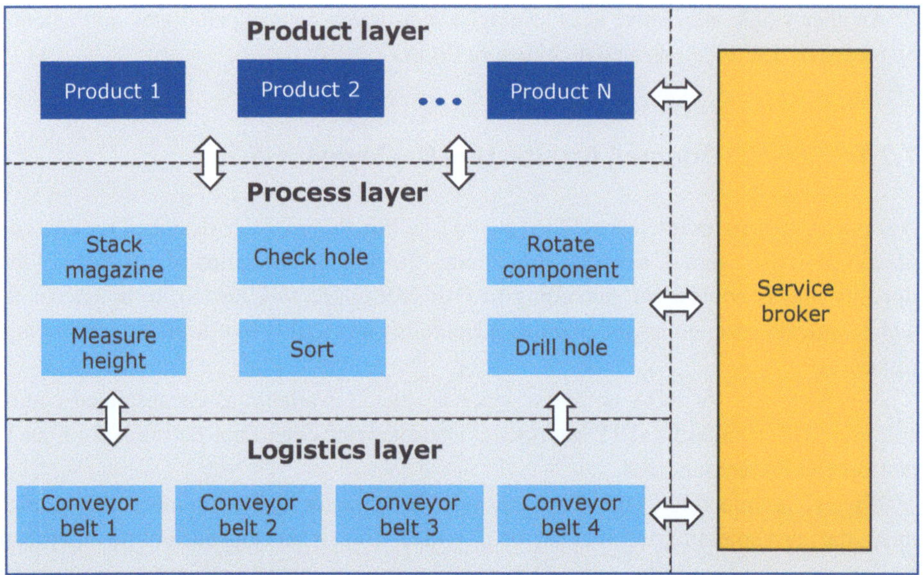

Fig. 7.4 Layered model of SOA services for the automation of the model facility

Three layers are introduced to which the functionalities of the subsystems are allocated, i.e. the product, the manufacturing modules and the logistics modules.

The implementation of a service-oriented architecture (SOA) for the control of the modular manufacturing system requires the "product layer", the "process layer" and the "logistics layer", in which functionalities are provided in the sense of callable services.

Figure 7.4 depicts the services that can be obtained from the system. As described above, the service broker acts as a central information hub with which all parties involved can communicate. A product "n" calls a service, which is then executed by the respective subsystem, i.e. a production and logistics module. This means that product "n", for example a blank, organizes its own processing by calling the appropriate services. These services are then executed in the process layer by an appropriate subsystem, i.e. one of the plant's production modules. Prior to this, the component checks whether the product "n" is already in the appropriate position and, if not, calls upon the logistics layer to organize transport via a conveyor belt.

Before each service call, communication takes place between the user of the service and the service provider via the service broker.

The SOA presented here allows the simple addition and removal of devices and their functionalities within the layered model. This does not require significant changes to control programming. Instead, we only need to adjust the information regarding functionality and the manner of provision in terms of defining the service within the layered architecture.

Consequently, subsystems can be easily redefined, exchanged or removed.

7.2.2   Assessment of the SOA

For the example of a modular production system, a SWOT analysis can identify strengths, weaknesses, opportunities, and threats in terms of internal and external impacts.

Figure 7.5 illustrates the various aspects of a SWOT analysis.

The benefits of SOA can quickly become significant, especially for medium to large enterprises with interconnected automation systems. This is because it is easy to replace or expand equipment with components from different vendors, thus enhancing interoperability and flexibility. In addition, operational processes are easy to monitor, ensuring high quality through data transparency to allow optimal operational control. Despite these benefits, the initial implementation effort and the lack of clear service interface specifications are potential drawbacks.

For example, in the endeavor to achieve a "common language" for describing the respective capabilities, the introduction of service-oriented architectures poses new challenges for data modeling. In particular, the software structures of SOA in conjunction with service definitions increase the complexity compared to the straightforward methods of cyclical network control with PLCs, which are less elegant but much easier to develop and maintain.

For this reason, a cost-benefit analysis should perform a critical examination of the implementation of an SOA in advance by means of a cost-benefit analysis.

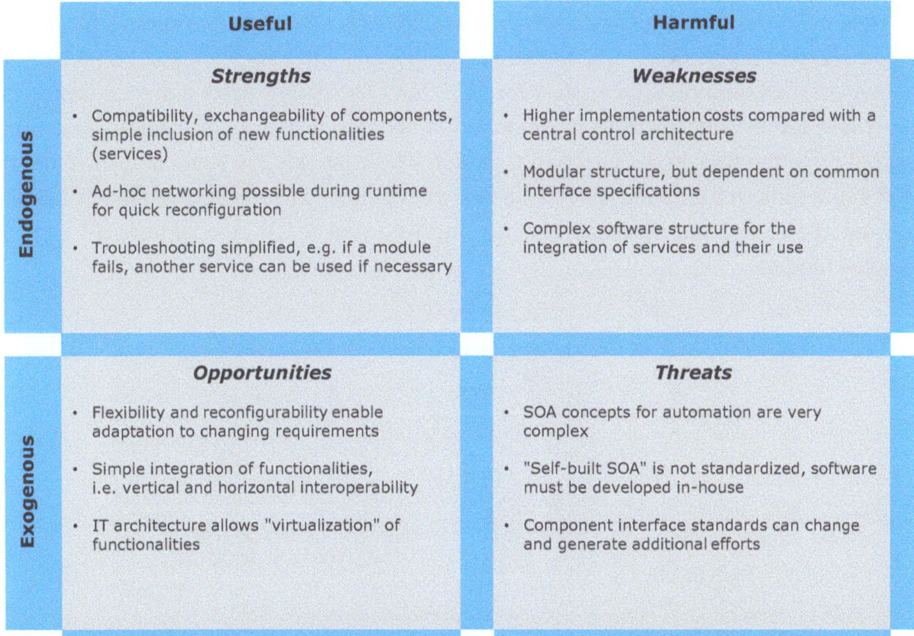

Fig. 7.5   SWOT analysis for the application of an SOA to a modular manufacturing system

7.3　Open Platform Communications Unified Architecture (OPC UA)

The Open Platform Communications Foundation (OPC) has established a standard for service-oriented architecture known as Unified Architecture (UA). This standard—OPC UA (2008)—has since defined a platform-independent architecture for automated systems which accommodates different industries and competitors. The purpose is to facilitate a seamless flow of information between automation components from different vendors. The standard and its protocols and information models are maintained and developed by the OPC Foundation and its industrial partners.

7.3.1　Concept of OPC UA

OPC UA is a platform- and vendor-independent standard for vertical and horizontal integration which enables interoperability between systems and components. It emphasizes the semantic description of information flows, data, and services for control, monitoring, and diagnostics. OPC UA provides services and communication interfaces for vertical and horizontal integration that are aligned with the Service Oriented Architecture (SOA) paradigm and specifically defined to ensure the automation of production across multiple industries. The architecture defines a hierarchy of information models that allows the inclusion of different industry segments underpinned by a basic definition.

Figure 7.6 illustrates the basic structure of OPC UA, which defines protocols and services for the integration of automation components. It also defines comprehensive metamodels and information models for OPC UA. The definition of a metamodel forms the basis for the construction of information models that are consistently structured according to the framework of the metamodel. These models are then complemented by domain-specific information models known as "companion information models" which are tailored to specific industries and are further expanded by the addition of vendor-specific models.

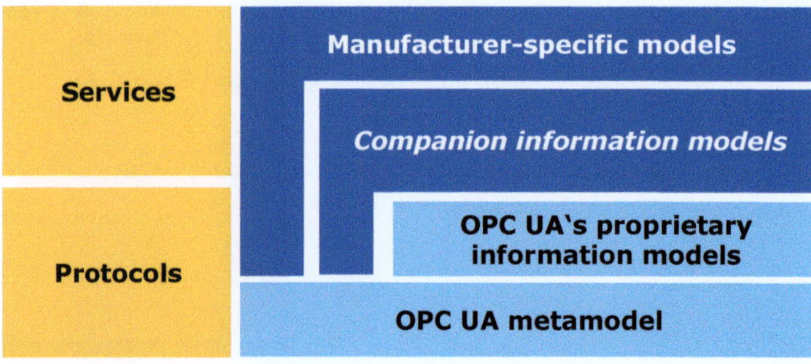

Fig. 7.6　The OPC UA concept

This approach provides a modeling framework that can be adapted to each domain and customized further by providers of automation components.

7.3.2   What Can OPC UA Do?

OPC UA enables the standardization of a service-oriented model for both services and interfaces. It improves the interoperability and scalability of production automation systems and is tailored to the needs of different industries. The standard is suitable for overarching solutions requiring control and coordination in horizontal and vertical integration contexts. For instance, triggering remote management commands for an industrial facility using a tablet connected to a server is easily accomplished with OPC UA. Predefined concepts for use cases of this kind are available in OPC UA.

However, OPC UA requires a sophisticated framework with a high demand for computing resources. The number of services in OPC UA can become very big in larger installations, resulting in considerable complexity. Implementing OPC UA for use in smaller facilities is relatively complicated due to the framework. While the service structure may be straightforward, using the OPC UA framework still requires significant effort compared to the benefits.

In general, the architecture requires stable communication between participants which is accompanied by significant power consumption. In addition, there are substantial licensing fees. Figure 7.7 shows the SWOT analysis of OPC UA.

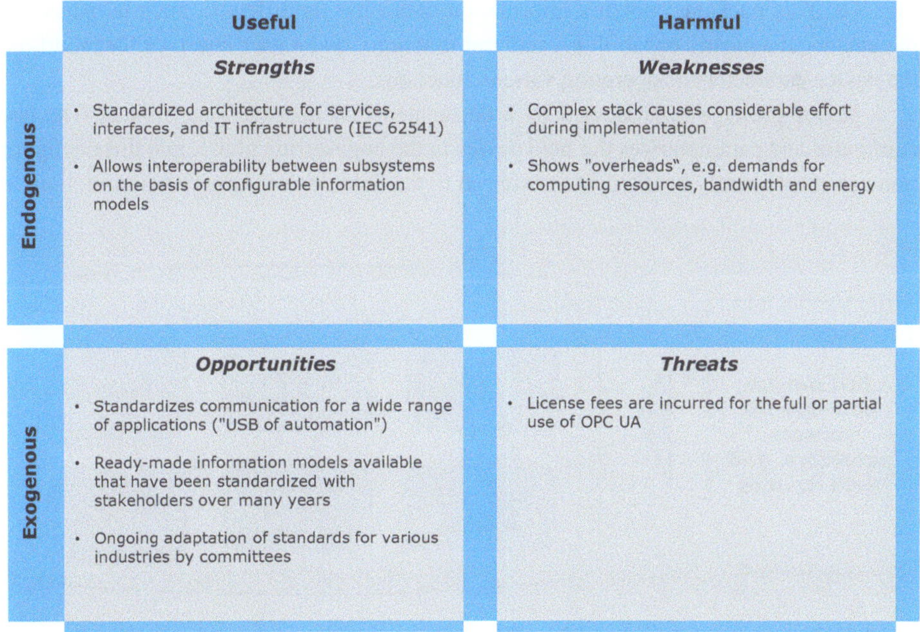

Fig. 7.7 SWOT analysis of OPC UA

7.4   Other Interoperability and IT Integration Concepts for Automation

Many other integration approaches and initiatives aim to achieve interoperability in the different domains. This chapter provides an overview of some standardization approaches and concepts that can be exploited for practical integration. It provides a high-level view of approaches and can be used as a starting point for further research.

This chapter provides an overview of Field Device Integration (FDI) and Module Type Package (MTP). FDI refers to the automatic integration of field devices whereas MTP refers to the integration of automated components in manufacturing and supervisory control. The NAMUR Open Architecture (NOA) is introduced to provide information coming from various levels of the automation pyramid.

Finally, these approaches are subjected to a comprehensive SWOT analysis and evaluation.

7.4.1   *Field Device Integration* (FDI)

Field Device Integration (FDI) is the integration of field devices. It is based on the standard DIN EN IEC 62769 (2021). This standard is a consolidation of previous approaches to device integration, i.e. the Electronic Device Description Language (EDDL) and the Field Device Tool (FDT).

A field device type is described in the form of an FDI device package as shown in Fig. 7.8. This package contains information about the field device, such as adjustable parameters, diagnostic capabilities, self-test functions, and a user interface for modifying the device parameters or operating various functions.

A device package is processed by a designated FDI host (service broker). The user configures and parameterizes the field device in the engineering phase, and this configuration can subsequently be directly transferred to the field device during commissioning. In

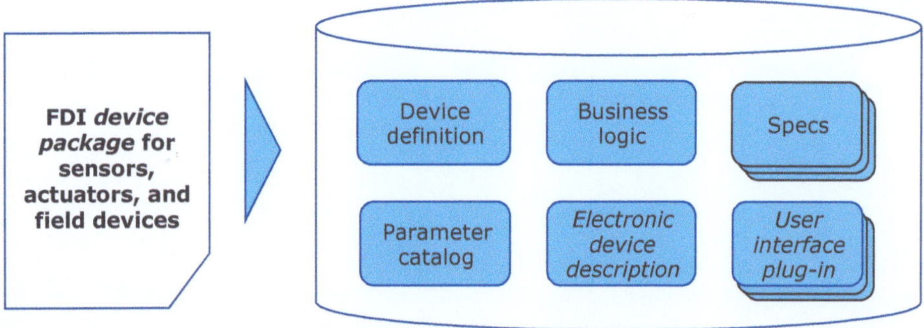

Fig. 7.8 Overview of the structure of the FDI device package

this way, a service broker represents the field devices as a service provider and makes their functions available via OPC UA.

To meet the operational requirements of a modular production system, FDI specifies a vendor-independent OPC UA interface and information model. The IT architecture of FDI follows the Service Oriented Architecture paradigm to meet the operational requirements of a modular production system.

7.4.2   *Module Type Package* (MTP)

The concept of Module Type Packages (MTP) originated in the process industry. It is used to automatically integrate automated manufacturing modules into a larger system. An MTP interfaces with alarm management, human-machine interface, system diagnostics, process control, and other system functions. Multiple modular process units can be assembled to create a process facility that is interconnected in terms of hardware. MTPs with integrated control are developed in the context of automation system engineering. These process equipment assemblies represent the process infrastructure of a modular facility and have the following characteristics:

– Interfaces are standardized and formally described.
– Modular process units can consist of one or more units.
– Production modules have extensive automation and safety autonomy, i.e. their functionality is encapsulated to allow them to operate autonomously.

Commercially available Manufacturing Execution Systems (MES) support the MTP standard by receiving and processing the information.

The guideline for MTP according to VDI/VDE/NAMUR 2658 (2019 to 2022) formally describes the interfaces and functions of the automation technology of a module. The MTP provider defines a "manifest file" that is modeled in Automation Markup Language (AutomationML). The MTP includes properties for services or information for operator displays in a machine-readable format that can be processed automatically.

To connect information flows and establish communication, MTP files for process control above the SCADA level are linked within the MES to anchor the modules. For example, information contained in the MTP file about the operator displays of each module can be automatically generated there, thus eliminating the need for manual construction and linking to process variables.

This concept also embraces the service-oriented paradigm and uses the OPC-UA standard for communication.

7.4.3   The NAMUR Open Architecture (NOA) for Vertical Integration

The NAMUR Open Architecture (2021) also referred to as NOA is an approach that takes services and data from field devices and supplies them to higher-level systems within the automation pyramid. Vertical integration is especially relevant in the process industry due to its complex plant structures. Consequently, NAMUR—the interest group for automation technology in the process industry—has initiated a joint effort with users of automation technology and digitalization in order to address this complex issue.

The NOA concept was developed to address the challenge of limited data connectivity in higher-level systems, especially between the field and control levels. As field-level sensors and actuators become more intelligent through digitalization, they generate a variety of measurements and status messages as well as vital data and identification data that could be processed at higher levels. However, due to the hierarchical architecture and the associated high integration effort, many data points remain unused and are often obstructed by proprietary interfaces and protocols. The NOA concept helps to transfer this data stream to higher-level systems, providing several advantages such as improved control monitoring or new opportunities for improving plant performance by means of data analysis.

NOA is an IT architecture that enables monitoring and optimization capabilities vertically along the automation pyramid with little effort. With the NOA architecture, a field device no longer has to be manually integrated into an existing IT coordination and monitoring structure of a facility but can communicate with other applications in the hierarchy via a parallel second communication channel. This allows the preservation of established industrial automation structures that have developed over many years.

The conceptual idea of NOA is illustrated in Fig. 7.9. It is based on the use of standard interfaces and protocols in conjunction with an information model that enables a com-

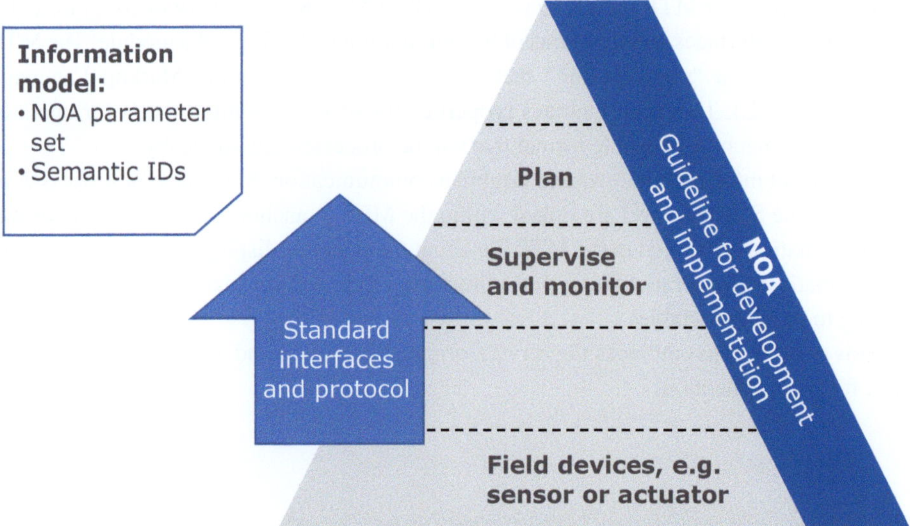

Fig. 7.9 Conceptual idea of the NOA architecture (NAMUR Open Architecture)

munication channel parallel to the automation pyramid. The information model includes an NOA parameter set and semantic IDs that describe the syntax and semantics of data exchange between levels. The parameter description is based on the standardized Process Automation—Device Information Model (PA-DIM), which in turn is based on OPC UA and specifically designed for data exchange between devices in process automation. By applying the NOA development and implementation guidelines consistently, information from the field device level can be transferred to higher levels for monitoring and planning using standard interfaces and protocols.

However, NOA is not a final standard but rather a concept that comes with its own set of challenges. The security of data transmission is a critical concern as the data must be purposefully transmitted, and NOA should never be misused as a means of organizing attacks on the facility through this side channel. A concept like a "data diode" that would ensure that data flows in the intended direction only would be a desirable notion. Still, implementing it is challenging due to the complexity of the software realization.

7.4.4   What Other Attempts at Interoperability Exist?

The field of IT architectures for interoperability is the subject of numerous research efforts, standardization initiatives, and industrial development processes. There are a number of other approaches that are impacting future IT architectures in production automation. The success of approaches to interoperability depends not just on technologies but also on economic and strategic considerations.

To explore these issues further, Table 7.1 provides a starting point for the further exploration of these issues by highlighting a number of successful initiatives.

Table 7.1   Other IT architectures for a range of application scenarios

BaSys	The BaSys4 reference architecture is an overarching term for so-called middleware and describes the entire platform on which communication takes place.	www.basys40.de https://projects.eclipse.org/ projects/dt.basyx
AAS of IDTA	The asset administration Shell (AAS) enables interfaces to the digital twin, which connects physical industrial products with the digital world.	industrialdigitaltwin.org
Open process automation	Users and providers collaborate to create a standards-based, open, secure, and interoperable process control architecture / open system architecture	www.opengroup.org/forum/ open-process-automation- forum
IIRA	Architecture for the development of interoperable, industrial Internet of Things systems for applications in the industrial domains	www.iiconsortium.org

The technical issues are complex and can only be thoroughly addressed by means of industrial projects under real operating conditions. Non-technical considerations such as dependencies on IT architecture vendors, emerging system complexities, the interchangeability of system components etc. also influence factors that should be assessed economically and politically. For this reason, several approaches, standardizations, and collaborative initiatives are expected to develop this multifaceted topic further.

7.5   Prompts for Reflection

In principle, interoperability approaches promise the platform-independent integration of automation components. Implementing reconfigurable automation in component integration could automate part of the updates, leading to a significant reduction in manual effort and cost.

Each of the interoperability approaches presented here is based on a semantic description of data within the confines of an information model. However, these approaches focus on different application areas and consider different levels of automation. As a result, the approaches have their differences but generally share similar goals of interoperability and seamless data and information processing.

Standardizing IT makes it easier for operators to manage automation systems and reduces the need for custom software. However, the adoption of IT architectures initially introduces increased complexity and effort due to additional IT infrastructure. Although process management is simplified for complex systems, complex implementations are required here.

Weighing the real benefits of interoperability is critical. What new capabilities does interoperability really provide? How does increased flexibility, improved scalability or simplified system management translate into tangible value? Which cost optimization opportunities can be quantified? These questions are countered by the implementation effort, which bears significant risks due to the high complexity of the software and potential compatibility issues.

The standardization effect in modules also leads to increased competition among component manufacturers. On the one hand, they can offer plant operators greater choice and the potential for cost optimization. On the other hand, though, there is considerable pressure because the components are easily interchangeable and there is a risk of losing markets to competitors. This explains why norms and standards are implemented slowly or even circumvented altogether.

Develop hypothetical situations that could arise in the future due to the use of the new IT architectures.

The following questions are relevant in this connection:

a. What are the advantages and disadvantages for the operating enterprises and the vendors of automation technology? What are the possible extreme scenarios (worst case and best case)?
b. What are the strategic implications for the implementation of new automation technology and the operation of systems? Which influencing factors drive a scenario in one direction or the other?

In particular, consider the issues of simplification and complexity in the deployment of IT and discuss the potential business opportunities for the companies concerned.

Further Reading

Bittorf, L.; Beisswenger, L.; Erdmann, D.; Lorenz, J.; Klose, A.; Lange, H.; Urbas, L.; Markaj, A.; Fay, A.: **Upcoming domains for the MTP and an evaluation of its usability for electrolysis.** 27th IEEE International Conference on Emerging Technologies and Factory Automation (ETFA), Stuttgart, Germany, 2022. https://doi.org/10.1109/ETFA52439.2022.9921280

Colombo, A. W.; Karnouskos, S.; Mendes, J. M.: **Factory of the future: A service-oriented system of modular, dynamic reconfigurable and collaborative systems.** In: Benyoucef, L.; Grabot, B. (eds.): Artificial intelligence techniques for networked manufacturing enterprises management. Springer Series in Advanced Manufacturing. Springer, London 2010. https://doi.org/10.1007/978-1-84996-119-6_15

Köcher, A.; Beers, L.; Fay, A.: **A mapping approach to convert MTPs into a capability and skill ontology.** 27th IEEE International Conference on Emerging Technologies and Factory Automation (ETFA), Stuttgart, Germany, 2022. https://doi.org/10.1109/ETFA52439.2022.9921639

MacKenzie, C. M.; Laskey, K.; McCabe, F.; Brown P. F.; Metz, R.: **Reference model for service oriented architecture 1.0. OASIS Standard.** OASIS Open, 2006.

Tauchnitz, T. (Hrsg.): **MTP automation of modular plants.** Vulkan-Verlag, 2022

References

DIN EN IEC 62769: **Field device integration (FDI).** Teil 1 bis 4, Beuth-Verlag, 2021 bis 2023. https://doi.org/10.31030/3364570

DIN SPEC 16593-1:2018-04: **Reference Model for Industrie 4.0 Service Architectures - Part 1: Basic Concepts of an Interaction-based Architecture.** Beuth-Verlag, 2018. https://doi.org/10.31030/2838942

NAMUR Open Architecture: **NOA Information Modell**, No. 176, (in German) NAMUR- User Association of Automation Technology in Process Industries., Mai 2021

OPC UA: **Unified Architecture.** 2008. https://opcfoundation.org/about/opc-technologies/opc-ua/

VDI/VDE/NAMUR 2658: **Automation engineering of modular facilities in the process industry.** Part 1 to 3, (Draft in German) Beuth-Verlag, 2019 bis 2022

Case Study: Automotive IT Today and in the Future

8

Abstract

This case study presents an overview of the automotive IT domain and discusses potential technology strategies for the future. With the advent of connected and autonomous driving, many questions arise regarding the evolution of software, electrical/electronic systems, and IT. This case study explores the role which long-established and future IT architectures can play.

- Which technological trends are driving change in today's automotive IT architectures?
- How can solutions be found, and which issues need to be considered?
- How can technological considerations be integrated with strategic aspects, and which factors are worthy of attention?

To answer these questions and provide an insight into the complex world of automotive IT, this case study provides a brief overview of the state of the art and the technology landscape.

The automotive industry is undergoing several technological advances in the field of connected and autonomous driving, and their innovative potential is expected to serve as a benchmark for progress in and around vehicles in the coming years. In the future, vehicles will possess software capabilities that can access information from other vehicles and infrastructure. Future vehicles will become ubiquitous systems, offering functions such as fleet management, car sharing, comfort features, and autonomous driving capabilities. It is

M. Weyrich, *Industrial Automation and Information Technology*,
https://doi.org/10.1007/978-3-662-69243-1_8

conceivable that the market success of new vehicles will be determined by software-based capabilities of this kind as manufacturing quality and powertrain technology have largely innovated as far as they can and are no longer perceived as indicators of technical progress.

This case study on automotive IT discusses future software and communication architectures and their potential components. It is evident that, although in-vehicle software complexity is currently managed by functional, standardized IT system architectures, these solutions have become unworkable in practice, making it difficult to expand and innovate any further.

Implementing software-based developments in vehicles is already challenging because existing IT architectures are struggling to cope with the growing complexity—and this complexity is expected to increase. In turn, new requirements for software quality and reliability are emerging. To enable connected and autonomous driving, it is necessary to modify existing IT architectures and software concepts and thus promote future developments.

Within the scope of this case study, it is impossible to formulate a comprehensive future concept for practical application since the technical solutions to be expected are incredibly diverse. In view of the immense costs associated with technical implementation, these changes are not solely a matter of technical feasibility, but also involve economic interests as new technologies must prove their usefulness in order to gain acceptance. And then there is the question of industrial policy, the impact of which is hard to forecast.

Nevertheless, it is possible to sketch out several developments, outline various approaches to solutions, and illuminate a number of strategic aspects in general terms. However, it is essential to interpret the case study as an illustration which reveals current challenges in a key area of automation technology and outlines the impact of solutions without claiming to fully capture the complex realities, particularly considering the rapid evolution of this domain.

8.1   Technological Advances

The vehicle of the future will evolve into a networked, automated cyber-physical system equipped with extensive sensor arrays, communication technologies and advanced computing capabilities. Currently, Original Equipment Manufacturers (OEMs) and their suppliers are challenged by innovative technological approaches as the proportion of IT and software in the value chain has increased dramatically.

In addition, foreseeable technological trends are driving the evolution of existing vehicle IT architectures:

- There is an urgent need for a forward-looking IT architecture to shape cockpit design in the future by integrating navigation functions, infotainment, and possibly also applications for specific tasks such as maintenance, fleet management etc., with the aim of supporting the occupants in an optimal fashion.

- The networking of control units within the vehicle and with the infrastructure promises to increase the information available to the driver, enabling features such as warnings indicating traffic congestion, hazardous situations, or accidents.
- Autonomous driving requires advanced system capabilities in perception and decision-making, thus liberating the driver from routine tasks or potentially eliminating the need for manual driving in the future.
- There is potential for software updates during operation which range from over-the-air updates to continuous development and deployment loops. At the same time, a comprehensive information feedback loop is created to link vehicles in service back to the OEM's development department, enabling entirely new development processes.

In response, vehicle manufacturers and the IT industry are now developing novel IT architectures that focus on implementing the above capabilities. In this context, a number of anticipated technological trends are presented below.

8.1.1 Trend: Interconnection of Software-Defined Vehicles

Connecting vehicles to each other, to the outside world, and to a centralized cloud enables many new capabilities. This extensive availability of networking and data can also be used to coordinate systems or alliances of systems. Today, the use of navigation applications simplifies route planning and allows traffic jams and dangerous areas to be avoided. It is conceivable that enhanced mobility services such as optimized maintenance and ride-hailing services will become available in the future.

Several data loops will be created here (see Fig. 8.1).

Communication can be between the vehicle and everything else (Car-to-X), between vehicles (Car-to-Car, C2C), or between the vehicle and the infrastructure (Car-to-Infrastructure). In this way, vehicles can for example inform each other about traffic congestion or accidents and then coordinate their actions to ensure a safe traffic flow.

Connected car or mobility platforms are already enabling new forms of value creation based on this connectivity, such as car-sharing, ride-sharing, or intermodal mobility, in which different modes of transport are coordinated. Collecting this data and coordinating the various data centers is handled by an overarching cloud.

The vehicles themselves or a local data center (a so-called edge) can perform coordination for local fleet management and in particular real-time applications including many vehicles. In the process, data from nearby vehicles is collected and analyzed via 5G. Zaki (2018) expects the widespread availability of 5G to be a key technology for implementing these techniques.

The relationship with workshops is also changing as the manufacturer monitors vehicles in the field, allowing for the early initiation of predictive maintenance on the basis of operational data. Data collected in the cloud and stored in large databases (also known as data lakes) on the manufacturers' premises can be used for long-term monitoring and

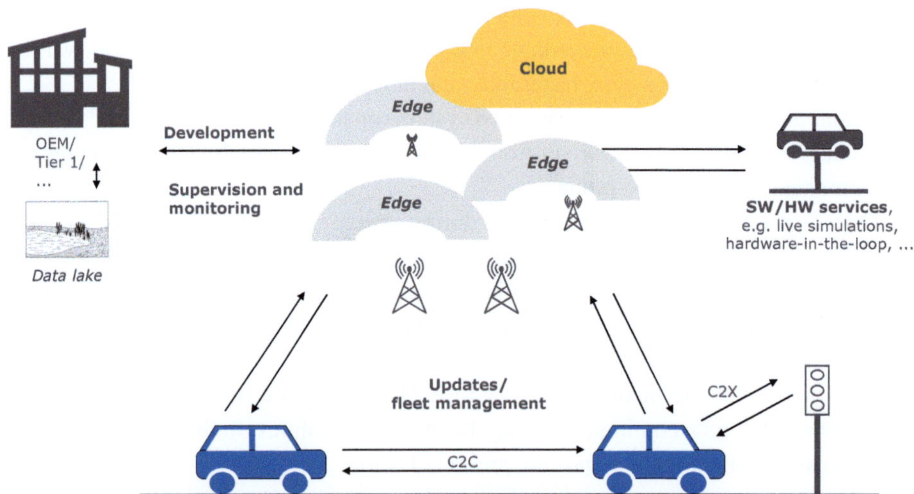

Fig. 8.1 Interconnection of software-defined vehicles in an end-to-end process between users and manufacturers

analysis. Developments of this kind in OEM and supplier practices cause the automotive sector to redefine systems engineering processes.

Over-the-air updates (OTA) can be used to improve a car's software-based functionality in the field without the need to visit the workshop. As these features are developed while the vehicle is in operation, new methods and techniques are needed to implement them. Currently, the emphasis is on the automated testing and deployment of software changes (continuous deployment) and the monitoring of the operational system to capture the effects of these changes and incorporate them into the development of new features. The resulting cycle of development and operations is the central feature of the so-called DevOps paradigm. Some examples of features that can benefit from an updating process of this kind are driver assistance systems, performance optimizations for battery-powered vehicles, and entertainment software for onboard computers.

8.1.2   Trend: Data Loop for the Continuous Development of Autonomous Systems

The connectivity of vehicles enables a continuous connection between engineering and the vehicle on the road. If this connectivity is reliable, data from the vehicle can be transmitted to a back end in real time for further processing. The data loop allows the real-time transmission of sensor and actuator data to a remote back end, which can be a control center. The data transmitted can be viewed, processed, and analyzed in this control center. The data loop is considered as a key technology for the development and deployment of autonomous driving for the following reasons:

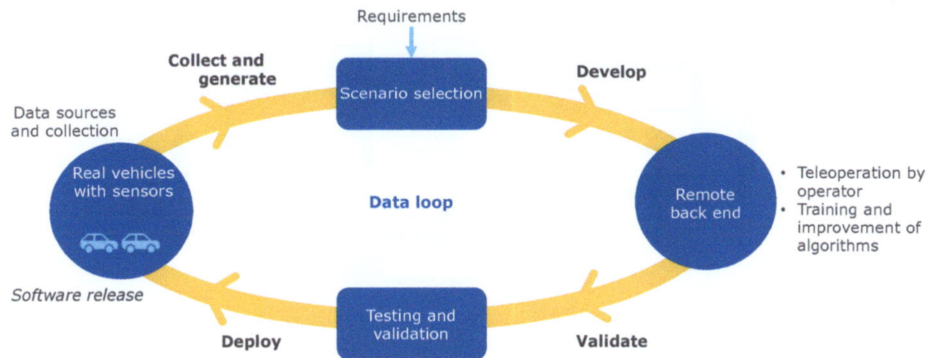

Fig. 8.2 Data loop, development cycle for autonomous systems

- In the development of AI controllers, targeted training with real and synthetic scenarios along with validation testing is crucial for the quality. The more real-world data is available, the better an AI system can be trained and validated.
- In the foreseeable future, it will be necessary to control the autonomous system remotely via teleoperation from a central control center when it encounters situations out of which it cannot maneuver automatically.

Figure 8.2 illustrates a data loop of this kind. It collects comprehensive information from the vehicle's sensors, receives it for further processing, and then transmits it to the development process.

Once data security issues have been addressed (including considerations of which potentially confidential information can be processed), the development of AI based on real-world situations can be pursued in the back end.

This also allows an operator to take control in real time. In this way, new algorithms and functions can be tested and validated in the background on real vehicles during ongoing operation.

8.1.3   Trend: New Technology Platform for Autonomous Driving

Autonomous driving requires an IT architecture able to conduct comprehensive mathematical calculations rather than relying on a multitude of small, decentralized control units as in the current practice. Autonomous and connected driving will require real-time operating systems and potentially new types of IT platforms to perform sensor processing, autonomy functions, and actuator control. Figure 8.3 schematically depicts the well-known robotics paradigm of "sense", "plan" and "act".

It is foreseeable that a wide range of sensors will be used to sense the environment in the future, and their information will subsequently contribute to the action planning of the autonomous vehicle.

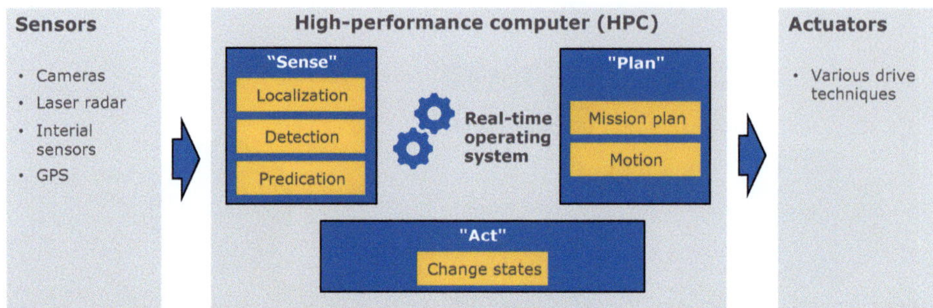

Fig. 8.3 Structure and functionalities of an automation system for autonomous driving

High-performance computing (HPC) for autonomous and connected driving also enables computing-intensive processing such as sensor fusion and advanced AI algorithms for action planning. The deployment of HPC makes it possible to implement solutions for autonomous and connected driving. At the same time, these HPC systems can take over the tasks of small ECUs and partially replace them.

8.2   An Important Baseline: The AUTOSAR Development and Integration Platform

Since the early 2000s, it has been recognized that vehicles will increasingly use electronic control units (ECUs) as the number of signals to be processed continues to grow. In response, the AUTomotive Open System Architecture (AUTOSAR) was established in 2003 to enable integration based on a unified standard. AUTOSAR provides a formalized description of software and hardware components and a runtime environment—the AUTOSAR stack—for hardware abstraction. This standard of the automotive industry provides uniform interfaces, interoperability, and modularity and is now well established with numerous automotive manufacturers.

However, the standardized system architecture is complex, resulting in many documents and a complex software stack that is resource-intensive to maintain, use, and extend. The promise of AUTOSAR is based on a layered architecture that formalizes the interactions between sensors and actuators in the vehicle. Figure 8.4 describes the layers and their tasks as standardized by AUTOSAR.

For example, suppose the left turn signal is to be activated. In this case, the runtime environment sends the control command via a standardized virtual function bus that coordinates the signal flow to the correct subsystem. Upon arrival, the AUTOSAR stack translates the command into actions executed by the hardware of the subsystem. The software accesses the correct address in the hardware via the runtime environment, the I/O abstraction layer, and a driver, and because of the hardware abstraction, software develop-

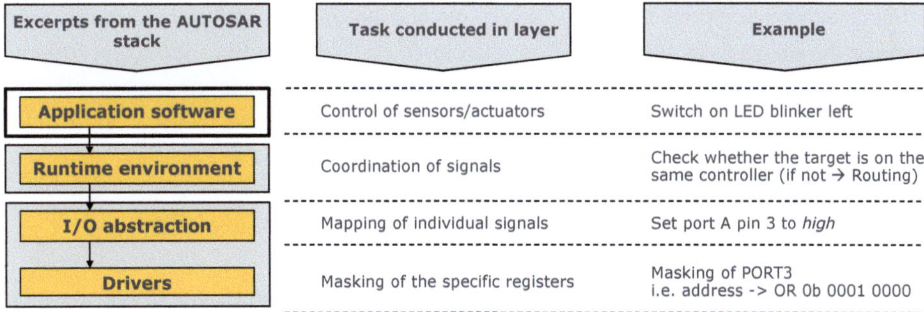

Fig. 8.4 Excerpt from the AUTOSAR reference architecture with special emphasis on the abstraction of input/output

ment does not need to know the details of the architecture. Thanks to its layered architecture, the software is not dependent on specific hardware.

Application software can be made more portable through the abstraction of the hardware input/output and the specific control. The AUTOSAR basic software package provides hardware abstraction and system software that allows ECU hardware to be exchanged without the need to modify the application software. By providing a standardized virtual function bus that coordinates the signal flow, AUTOSAR ensures interoperability. The runtime environment and the basic software are structured in the form of modules or function blocks. AUTOSAR can be viewed as a network operating system that integrates different ECUs, implements uniform interfaces, and ensures the use of a common basic platform and a common design methodology.

Although this standard has been agreed upon by the car manufacturers and implemented with the suppliers, the IT architecture of the software stack is large and monolithic, requiring significant effort for the use of AUTOSAR. In fact, because AUTOSAR involves a large number of contributors, changes are subject to extensive discussion and approval processes, leading to time-consuming adaptations.

With the evolution of the AUTOSAR adaptive standard, over-the-air updates are now possible, thus enabling the implementation of advanced functionality in the future.

8.3   On the Way to Finding Solutions

Various questions about possible developments arise from the above technical examination of automotive IT. It is obvious that these technologies are being driven by customer demand and are increasingly being perceived as key technologies for the modern car of the future.

However, what should the strategy for future technology configurations be? What specific paths can be taken? How should things be implemented, and which subjects have priority? Which interests drive development, and which may obstruct it?

Several considerations have to be taken into account in the search for solutions. These include the question as to which IT architectures could be used in the future. It is also important to consider the competitive position of established manufacturers and the true extent of added value achieved by software and IT in the automotive industry of the future.

8.3.1 Interviews with Experts on IT for the Future of Mobility

This makes it important to further elaborate on the trends by giving the opinions of experts and making the perspectives of specific interest groups visible. According to Essam Khamis (2022), the Delphi method -, a structured questioning technique for future research—is used for this purpose. According to this method, experts are asked for their opinions, and their assessments are then elaborated systematically. However, the selection of questions and experts has a significant impact on the results, and the information gained using the method should not be overestimated as it is more a structured representation of the opinions of a typically quite small group on a particular topic. Nevertheless, the method is frequently used to set priorities in development teams, although assessments can change quickly, for example when competitors implement technologies successfully.

Table 8.1 shows some expert assessments of the future of mobility.

A clear assessment of these issues that could be used to set priorities seems not to have emerged yet. The specific value is not defined well, and predicting customer acceptance is difficult even if the utility of a feature is rated as higher or lower.

Due to the significant success of individual OEMs in deploying vehicle computers—i.e. powerful HPC for managing IT-related operations—and as a result of the aforementioned centralization of control units, HPC is expected to become a fundamental

Table 8.1 Illuminating the road ahead: expert assessments of the mobility of the future

Subject	Assessment	Added value	Efforts
New **IT architectures** based on HPC	More centralized computing power is a key prerequisite for new IT architectures		
Enablement of continuous **software updates**	Over-the-air to DevOps, CI/CD can be very complex due to IT security		to
Deployment of advanced **software development** methods	Open-source software development is difficult due to the legal situation		
Networking of vehicles with each other and with the infrastructure	Scalable networks as well as edge and cloud computing solutions are a prerequisite		
Capabilities to enhance driving comfort at **SAE 3, 4 and 5** levels	Assistance systems for autonomous driving are perceived as milestones of innovation	to	

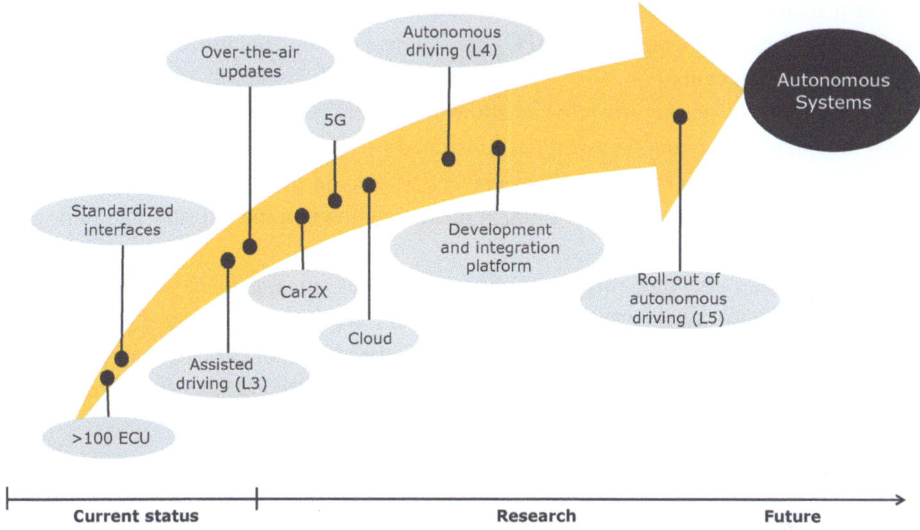

Fig. 8.5 Roadmap for "IT for automotive" today and in the future

requirement for functions such as autonomous driving, entertainment, navigation etc. in the future.

However, it remains quite uncertain whether certain autonomous driving capabilities should be based on constant connectivity to the cloud, allowing them to be realized cost-effectively there, or whether the associated connection to the customer would be perceived as surveillance.

The only way to further specify the course of development is to elaborate on the questions and include them in the development process. In light of the rather vague future topics, it is recommended to conduct applied research and support for start-ups in the seed stages of development. These measures can make innovations more tangible by implementing demonstrators, feasibility studies or prototypes.

Figure 8.5 is an attempt to use a roadmap to outline a possible pathway for development, at least in general terms, as well as potentially relevant technologies. Obviously, due to the multitude of different aspects to be considered, it is uncertain whether these technologies will emerge in the order depicted here.

8.3.2   A Concrete Example

Once an approach based on the opinions of experts in the domain and possibly also on the observations of competitors emerges, development management can be used to develop it further.

For instance, if the consensus is to implement new IT or software architectures based on High-Performance Computing (HPC), this subject can be concretized with the help of prototype developments.

Observations of market participants show that central vehicle computers offer advantages in both software and hardware, as illustrated by Lock et al. (2020). HPC allows the use of IT architectures based on client-server concepts in the vehicle, thus enabling the application of many software development methods known in the industry and improving complexity management.

HPC facilitates real-time programming, multi-process management or automatic code generation. In turn, only a few central computers are needed to manage all the key software-related processes. An additional benefit is the reduction of fieldbus systems in the vehicle, something which, according to Kume (2020), leads to cost advantages in production thanks to less assembly and less wiring.

Here is a comparison of the advantages and disadvantages:

Advantages:

- The use of virtualization technologies improves the manageability of software complexity and simplifies software reuse.
- Innovations in software create new competition among suppliers of software modules.
- In the hardware domain, the ability to exchange actuators and sensors via standardized interfaces is established.
- The implementation of autonomous driving requirements becomes easier with sufficient computing power from HPC.

However, these benefits also come with drawbacks or risks:

- Existing supply structures and business relationships may change to the detriment of suppliers.
- New legal issues arise, such as the liability for the correct functioning of software components.
- Knowledge development and the availability of trained experts are required for new software stacks.
- Standardized development processes are not yet established, so that significant process changes based on new technologies and workflows are required.

There are three basic options when weighing the pros and cons:

OEMs can either maintain their existing decentralized approach, embrace the new philosophy of centralizing computing resources in HPC or choose the middle ground: a gradual transition and hybrid operation involving both HPC and fieldbus/ECU.

However, attention should also be paid to implementation with development partners as success or failure in the automotive sector in particular can depend significantly on the supplier ecosystem.

8.4   **Strategic Prospects and Interests**

Planning for new developments can be challenging due to uncertain assessments of options and risks. It is also important to consider the interests of multiple stakeholders. It is even more important to be aware that what may seem like a good solution today may soon become outdated due to future developments.

This evolution also raises a number of issues which are outside the realm of technical considerations yet still have a profound impact on technology design. Ultimately, new developments need to prove themselves quickly in the marketplace and—in the ideal case—create tangible value in the first phase of implementation.

A comprehensive strategy for positioning oneself for future developments often cannot be implemented by means of isolated technology enhancements. This is particularly important when it comes to building large-scale IT infrastructures in which customer value, acceptance, and ultimately business success require a great amount of entrepreneurial insight.

Figure 8.6 illustrates the perspectives that should be considered when building a strategy.

Innovations must be created from the customer's point of view, economic aspects of the business need to be considered, the company's own capabilities have to be considered, and competitors have to be kept in mind. The success of new technology in the marketplace must be the result of an examination of a wide range of issues in addition to technical considerations.

Consequently, the technological evolution toward autonomous and connected vehicles is full of challenges, and ignoring critical matters can have far-reaching consequences for the actions of original equipment manufacturers (OEMs) and their suppliers.

Fig. 8.6 Strategic aspects with regard to the introduction of new technologies

8.4.1   Strategic Perspective: Innovations for the Customer

Understanding what solutions customers will demand in the coming years is critical for future success and acceptance. However, customer preferences and the value of technology-driven innovation are often difficult or virtually impossible to predict. Customers may have difficulties envisioning how new technologies will function in the application, or they may overestimate partial features, only to lose interest when these become available.

Henry Ford is quoted as having said, "If I had asked people what they wanted, they would have said faster horses." This statement reflects the difficulty of identifying viable future advertising promises with a narrow market focus during a period of disruptive technology development.

In the realm of connected and autonomous driving, many aspects of value and user acceptance remain unclear. How customers will perceive the value of solutions for specific vehicle functions remains to be seen. In addition to the excitement that new features of connected and autonomous driving or entertainment systems can generate, various foreseeable challenges could diminish their value. These challenges include legal and organizational frameworks as well as societal acceptance of machine and IT security, which is viewed rather differently in certain parts of the world.

8.4.2   Strategic Perspective: Opportunities for the Business

How can the long-term transformation of the automotive industry be commercially successful? How will value be generated in the future, and who would benefit if, for example, sports cars could only be driven on racetracks in the future, with everyday traffic being chauffeured by AI?

The current market leadership of car manufacturers allows them to allocate resources for developing new technologies. However, securing access to software and electronic components for new vehicle computers, for example, requires strategic partnerships. It is crucial to avoid a "lock-in" effect with only one supplier for key technologies in order to avoid excessive influence and loss of value creation.

Software and IT could significantly alter the OEMs' business model if IT-based services, software applications or an automotive operating system from third-party providers were to become established and product-determining. It is therefore essential to ensure that emerging customer needs for new capabilities in this area are given thorough consideration in order to mitigate the risk of an uncontrollable and detrimental shift in the business later on.

The challenge is obvious: established companies are hesitant with very large investments in the development of innovative future technologies for autonomous and connected driving because their business success is not yet clear. On the one hand, this cautious approach can be damaging if more innovative competitors get a chance to implement successful developments first. On the other hand, the first developments of connected, intelligent products to market have been commercial failures, suggesting that investing in innovation "at all costs" can have negative consequences if the market is not ready. The key is to introduce products into the market gradually in order to achieve commercial success in times of rapid technological development.

8.4.3 Strategic Perspective: Inside View of Automotive Companies

Successful companies typically have large teams dedicated to designing today's products. However, when it comes to disruptive innovation, either employees will have to adapt their skills or other individuals will have to master the suddenly essential key technologies for designing future products.

In terms of key technologies, existing automotive companies will need to anticipate how to deal with large-scale software systems, new vehicle operating systems, AI software, and scalable cloud/edge technologies. On the IT side, high-performance computing (HPC), sensors, and new bus systems are critical for the future.

It is foreseeable that automotive OEMs and suppliers will cover IT technology much more extensively in the future in order to carry out very large software developments. This will require significant investments in automotive software engineering processes and employee skills development. IT tools and methods need to be made available for development processes, and employees should be trained accordingly.

In any case, substantial changes to existing automotive companies are necessary in order to cope with the upcoming shift in competencies of many of the employees affected, as outlined by T-Systems (2020), for example. If this transformation is unsuccessful, value-added shares will inevitably be lost or will migrate from existing companies to others.

Partner companies from the IT sector as well as start-ups could be involved in order to overcome emerging competence deficits and resource bottlenecks. There are many ways to do this, one of then being to open up reference architectures as a prerequisite for IT integration or creating common platforms that could include open-source solutions.

Given new technologies, existing partnerships in the IT sector with software and electronics manufacturers will have to be reconsidered to ensure future-proof positioning.

8.4.4 Strategic Perspective: Competitors Seek New Opportunities

It is quite obvious that the future of mobility will be shaped by the collaboration between car manufacturers and major IT, software, and communications companies. The make-or-buy decision of the major players will have far-reaching consequences for the future if it results in key technologies falling into the hands of new actors. On the one hand, established companies may find it difficult to develop the necessary technologies within their existing structures.

A retrospective look at phases of disruptive technological change shows that new technology partners have the potential to transform the market in their favor. New entrants attempt to overtake established OEMs and suppliers with their software and IT expertise.

However, the question is whether these new companies are really so far ahead of the established companies that they cannot be caught up with later. It should also be considered that unpredictable risks may arise if existing supply chains, and potentially the entire value chain, are disrupted. From today's perspective, it is often very difficult to anticipate how the future value distribution will take shape as all of the perspectives outlined above will have some impact. Nevertheless, the history of many failures suggests that companies are well advised to take competition in and around new technologies very seriously and invest significant resources in the research and development of emerging technologies.

8.5   Prompts for Reflection

This case study of an evolving industry illustrates how new technologies can change the world. The Internet of Things and connected vehicles will bring about many changes, and this case study gives us an inkling of how these technologies can shape the future.

It is already clear that connected vehicles will bring significant change, even if the specific applications and capabilities are not yet fully understood. High-Performance Computing (HPC) in conjunction with data looping is considered as a key technology for autonomous driving. While HPC and data looping are not currently mandatory for the operation of conventional vehicles, they are an essential technological foundation for the development of autonomous vehicles. This raises the question as to how these technologies should be phased in.

Autonomous driving is an example of a technology with immense depth that needs to be mastered. It will take many years and significant investment to bring the demanding research and development work in this area to a level of maturity which leads to affordable products. The path to commercial success for this technology is currently vague and uncertain. Ensuring commercial success while advancing research and development in this area is a challenging task.

Different strategic approaches from different perspectives need to be developed and explored to develop viable future concepts for companies in the automotive industry.

Some central questions for future development are:

- Which technology segments should established OEMs and suppliers occupy, where should they invest in partnerships, and which segments should they abandon?
- What is the role of new competitors that excel in software and IT?
- How should IT technology platforms for autonomous driving be evaluated in this context?
- What do the new technologies mean for existing systems engineering processes and the expected complexity of software?

Conduct your own research to find up-to-date answers to these questions.

Further Reading

Ebert, Ch.; Hochstein, l.: **DevOps in practice**. IEEE Software, Vol. 40, no. 1, 2023, https://ieeexplore.ieee.org/document/9994066

Shahin, M.; Ali Babar, M.; Zhu, L.: **Continuous integration, delivery and deployment: A systematic review on approaches, tools, challenges and practices.** In: IEEE Access, vol. 5, pp. 3909–3943, 2017. https://doi.org/10.1109/ACCESS.2017.2685629

References

Essam Khamis, N. : **The Delphi Method as a morphological catalyst for foresight-oriented design research.** Diid — Disegno Industriale Industrial Design, (76), 12. 2022. https://doi.org/10.30682/diid7622j

Lock, A.; Tracey, N.; Zerfowski, D.: **Heading for new worlds - New EE architectures with vehicle computers generate new opportunities** (in German). Automobil-Elektronik No. 04-05, 2020

Kume, H.: **Tesla teardown finds electronics 6 years ahead of Toyota and VW.** NIKKEI Asia, 17.02.2020

T-Systems: **Electrics and Electronics in Vehicles - Why Modern Cars are Turning into High-Performance Computers** (in German). Automotive IT, Firmenschrift, 2020

Zaki, M.: **C-V2X is paving the way to 5G for autonomous driving** (in German). All-electronics. de, Hüthig-Verlag, 2018

How Does Automation Create Value?

9

Abstract

Automation technology does not have a purpose of its own, but rather aims to perform tasks and, in doing so, generate an increase in added value for the application.

For this reason, the focus of this chapter is on the question as to how automation technology has the ability to generate value in a meaningful way. The following questions will be the subject of a closer look:

- What is value creation through automation?
- What examples of value creation through automation can be identified?
- How can automation technology be methodically analyzed and designed with a focus on benefits and value creation?
- How can the capabilities of an automated system be aligned with a potential business perspective?

This chapter presents two examples of different methods that can be used to analyze and present the value of an automated product or system. Based on value analyses and promises, it is possible to estimate the added value of new technical capabilities and derive development priorities.

9.1   Generating Value

For decades now, the rigorous use of automation technology has been a catalyst for business improvement and a key driver for the productivity and efficiency of technical systems. The use of automation technology leads to the generation of added value for the application.

But how is the concept of value to be understood in the context of automation technology? VDI Guideline 2800 (2010) defines it as follows:

> A value is the relationship between the contribution of the (automated) function to the fulfillment of needs and the resources used to achieve that fulfillment.

The value of an automation function derives from its ability to make the technical process more efficient compared to the manual execution of the process. Added value also results from consistent quality, reproducible results, and reliable execution.

For example, welding can be performed either automatically by a robot or manually by a human. The benefits of automation technology are realized when the cost of using the robot is lower than the cost of employing a human. Additionally, it should be considered that the welding robot adds value by providing improved weld quality as it performs the weld more consistently and reproducibly compared to a human.

When we examine automated functions, the focus is often on a specific, delineated process step such as workpiece machining, automatic route planning, or temperature control in a heating system. In this context, value creation through automation technology can be defined as process improvements such as the more efficient manufacturing of devices like robots or machine tools, the optimized provision of raw materials from the supply chain to production or the accelerated development of systems.

To meet the demands of volatile markets, value-added processes should be able to respond flexibly and elastically to the current business situation. The use of automation and industrial information technology in production systems and product automation plays a key role in this. For example, data collection in logistics, and hence the optimization of transportation, is often impossible without automation technology.

Nowadays, the technical feasibility of automated functions does not play the deciding role when implementing automation technology. More important is the benefit to the application—a benefit which justifies the effort and adds value through automation.

9.2   Unleashing Value: Examples of Real-World Gains and Added Value Through Automation

The development paths of automation technology regarding the creation of value in industry extend over applications to be found in all industrial sectors and indicate typical directions to go in.

There continues to be a focus on improving operational efficiency, for example due to the capability to produce faster and in higher volumes through the deployment of robotics. The evaluation of the cost-benefit ratio of an automation system continues to be a key factor in its implementation.

Communication networks allow us to keep in touch with products after they are shipped. For example, software changes can be made to a product while it operates. This enables innovative business models and customer services by providing a direct link to customers and value chain partners.

New opportunities also arise from the data economy, which has become increasingly important in recent years given the need to coordinate and organize. For example, the traceability of CO_2 emissions and the circular economy can only be achieved in a meaningful way on the basis of an automated data economy.

The following sections highlight value creation opportunities that increase operational efficiency, consider novel forms of customer support, and generate added value through the collection, organization, and analysis of data.

9.2.1   Added Value Due to the Increase in Operational Efficiency

Manufacturing companies increasingly expect customizable "intelligent behavior" that adapts to the current situation and can potentially reshape it. The business of manufacturers demands more agile and rapid production adjustments based on the information available.

Figure 9.1 outlines the capability of advanced automation technology to automatically respond to changes in the manufacturing process through reconfiguration.

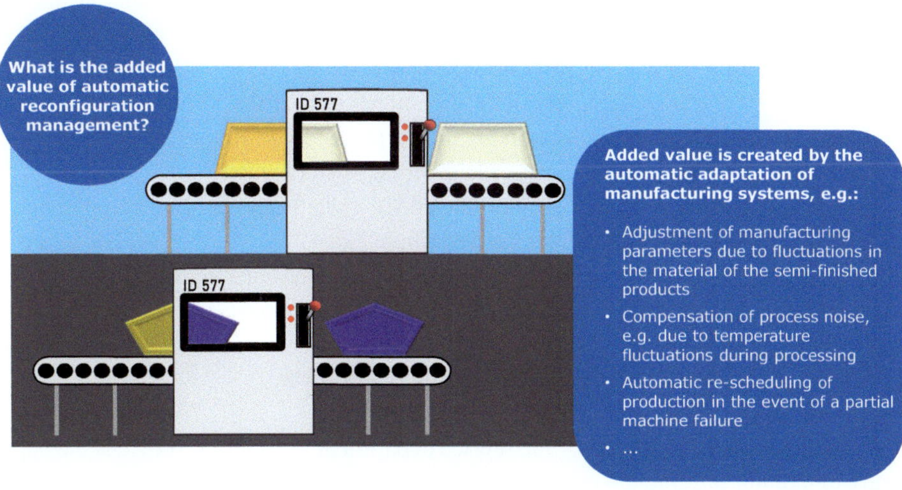

Fig. 9.1   Added value by automatically adapting manufacturing systems

Typically, there is a desire for a high degree of automation in order to minimize the need for personnel in the domain of production systems automation while at the same time being able to respond quickly to changes or external influences. The spectrum of relevant technical activities is broad. Research efforts range from the automatic diagnosis and self-calibration of the industrial facilities to the ability to assess and modify faults or alarms via remote maintenance and even to visionary concepts of self-adapting production.

Establishing effectiveness and efficiency in operations, and thus operating more cost-effectively than the competition, is the key benefit of industrial automation. It includes quality assurance, the avoidance of downtime, and the ability to implement necessary production changes with great speed and agility.

For example, the increasing need to realize small batch sizes and the difficulty inherent in predicting all of the future configurations of a system at the time of development lead to changing demands on a production system during its operation.

In many cases, however, these changes go beyond the pre-planned flexibility corridor of the production system.

This necessitates a rebuild, something which currently requires a high level of expertise, for example to continue to utilize parts of the existing facility and shorten the time-consuming and error-prone commissioning processes.

For this reason, value-added functions are required in the form of automation systems that can change by themselves and adapt quickly to changes by means of reconfiguration. This automatic change can be thought of as "automating the automated".

9.2.2 Added Value: Direct Link to the Automation Systems in Operation

It is becoming increasingly easy to connect directly to automation systems in the field using high-bandwidth communication networks. However, concerns about connection latency, IT reliability, and IT security must be addressed.

The potential of this connectivity innovation needs to be discussed as a direct communication link of this kind to products in the field represents a profound technological shift in automation as well as an associated change in the business models.

Figure 9.2 provides some insight into the added value in the field of teleoperation or software development based on DevOps (see Chap. 2).

In principle, the communication link with automation systems in the field can be used for multiple purposes as described in the following two examples:

a. **Added value: remote access and use of the digital twin.**

Remote access, for example by the service department in factories, provides the opportunity to monitor or remotely control machine states or operations on the basis of sensors and external sources of data and information.

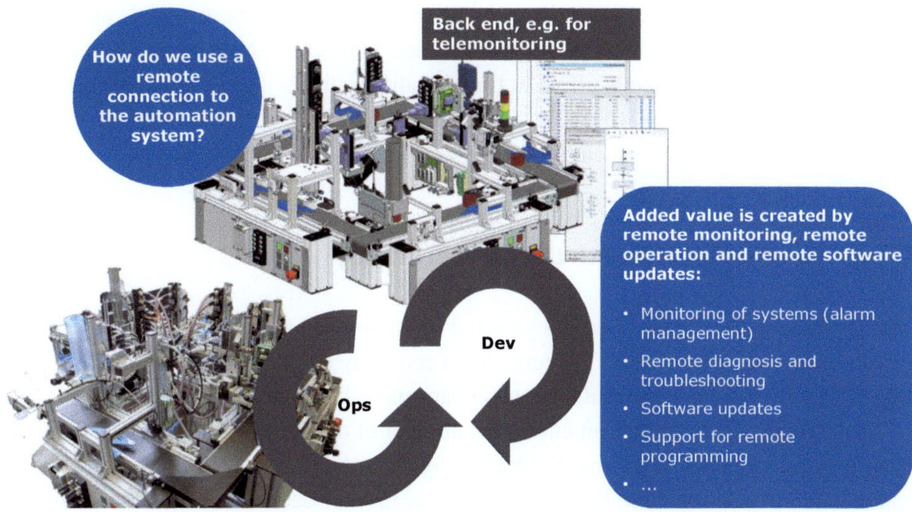

Fig. 9.2 Adding value to industrial production via teleoperation and DevOps

The ability to capture product usage and performance data in real time enables new service organization structures and processes. In addition, adaptability and agility in responding to customer needs is increased significantly because the situation on the ground can be assessed immediately.

For example, field service activities can be more targeted and thus more efficiently managed, or even made unnecessary, by alerting the personnel on site to improve a machine's performance or production quality.

As a result, teleoperation through trained and remotely guided personnel becomes possible on site.

A simulator can be operated as a digital twin in a remote control room, thus merging information with real machine data. In this manner, predictive maintenance of the digital twin can use diagnostic and analysis data from the real machine to detect and indicate problems at an early stage.

Some fundamental questions in this context include: How much functionality should be embedded in a product and how much should be in the cloud? What response times can realistically be achieved?

b. **Adding value: software changes for a back end far away.**

For software-intensive products, changes to the software can be carried out quickly "on the fly" while the product is running. These changes can be delivered over a wired connection or over the air, that is, via a wireless interface.

This allows for the instantaneous implementation of new features, rapid responses to change requests, and product customization—provided that the local network is available.

Due to the continuous communication link, entirely new business models are conceivable, possibly enabling new forms of maintenance contracts or even a pay-per-use model, i.e. billing based on the actual effort or functionality utilized.

However, the extent to which the end user actually benefits from frequent service or product upgrades needs to be clarified.

9.2.3   Added Value Due to the Data Economy

The leveraging of networked data is growing in importance and providing new opportunities for information discovery. The term "data economy" describes an economic system in which data is collected and analyzed using various technologies and platforms.

In the future, there will be an enormous amount of data collection going on. This real-time data provides value chains, product design, production, and customer service with immediate information from the field. The integration of data across individual functions through networked data will enable new capabilities and system functions.

Table 9.1 illustrates, based on two aspects in the railway sector, the added value that access to data can achieve and its utilization for functionalities at various system levels.

Access to data enables organizations to create new types of intelligent and connected products or services, transforming the customer experience and creating new business opportunities in the process.

Are users prepared to pay a higher price for these specific features, or do they just see them as nice things to have? At the end of the day, the success of automated functions depends on their acceptance by the users. As a result, it will take a combination of technology development, value analysis, and discourse between stakeholders to ensure that the users ultimately embrace them.

Connected automation systems raise economic and legal issues, too: How and by whom is added value created? How will the vast amounts of data be utilized and managed? Should the data of individual customers be used to offer improvements for everyone based on the information gained? How can we assess the risks of information aggregation from the viewpoint of security? Are there any regulatory frameworks? How are relationships

Table 9.1 Example for added value through networked data in the case of a train system

System	Data-based function	Added value
Bogies of the wagon (mechanical subsystem level)	Operating data from individual bogies and their subsystems can be recorded and evaluated online.	By comparing data, faults such as wear can be detected early and corrected proactively.
Train or rail system (overall system level)	An interactive travel assistant provides information about delays, alternative routes, and seat reservations.	Passengers can use an app to find their train connection and make reservations.

with traditional business partners being defined, and what changes are the latter experiencing in their industries?

Over and above the technical feasibility that will underpin the information economy of the future, countless business models and a number of legal and ethical issues still need to be explored and addressed.

9.2.4   Conclusion on Added Value and the Potential for Value Creation

Ultimately, the development will be influenced by technical feasibility on the one hand.

Automation and industrial information technology are obviously driving forces for new functionalities and capabilities that can create significant benefits and added value. However, given the new value adds, how should we assess their prospects of success? The following aspects may serve as a starting point.

Positive aspects:

- New automation capabilities that simplify the interaction with machines and products are emerging.
- New automation functions are ushering in improvements in our daily lives and the working environment, allowing resources to be saved and consumption to be monitored.
- New business models and service concepts set to change the industrial landscape are made possible.

Negative aspects:

- The automation of tasks will change the working environment: new tasks will come into being while others will become obsolete.
- New IT and machine security risks emerge as systems can be accessed remotely through networking.
- As systems become more complex, the predictable nature of individual functions will be reduced, causing challenges to reliability.

Ultimately, development is influenced by technical feasibility on the one hand. On the other hand, aspects such as the purchasing power of customers and value chain partners, the nature and intensity of the competition, the success strategy of market participants, and various socio-political considerations play a decisive role in implementation.

9.3   A Methodology for Value-Oriented System Design

A design methodology is required as a systematic approach that keeps value in mind and integrates value-oriented considerations into the system design of automation.

A value analysis is a systematic method for the structured and incremental development and analysis of functions. It considers various perspectives and influences, with a focus on identifying value-adding functions or capabilities.

9.3.1   Methodology of the Value Analysis

A value analysis methodology applicable to business and technology is defined in VDI Guideline 2800 (2010) with a corresponding work plan outlined in the DIN EN 12973 (2020) standard.

This approach focuses on the technical product, which is intended to provide a benefit or added value through its functions.

The individual steps of the methodology are oriented towards analyzing, describing, and developing. An assessment is made as to how and in which form an improvement of the existing functions or the development of new functions makes sense.

The emphasis is on the technical function that the product is supposed to fulfill or that is created by the interaction of product components.

The goal of this function is to satisfy a part of the user's needs. The methodology of value analysis describes the basic elements of the system and their interrelationships, analyzing and consolidating different perspectives.

A function-oriented approach in which the formulation of the problem or task is a prerequisite is a characteristic feature of the value analysis methodology.

The process of value analysis consists of steps which provide a preliminary examination of the initial situation and the problem that is supposed to be solved. Then, it systematically considers what a solution might look like. Ultimately, the best solution is selected on the basis of an evaluation.

Figure 9.3 shows an overview of the work steps in value-based system design.

The first step is to assess the situation. This involves defining, planning, and gathering relevant information and data.

A functional analysis is performed in the second phase. It includes the definition of a problem and a goal for the purpose of providing a clear definition of needs.

The goal of this functional analysis is to provide a description and a systematic representation, classification, and evaluation of functions and their relationships. This is achieved through the identification and formulation of user-related functions, evaluation criteria, and levels of satisfaction.

Subsequently, in the third phase, possible approaches to a solution are identified methodically by means of brainstorming.

Finally, in the fourth and last phase, a specific approach is selected and developed for implementation purposes. This approach is aligned with the capabilities of the product under consideration and is driven by the goal of optimizing product functionality in terms of value, resources, and effort.

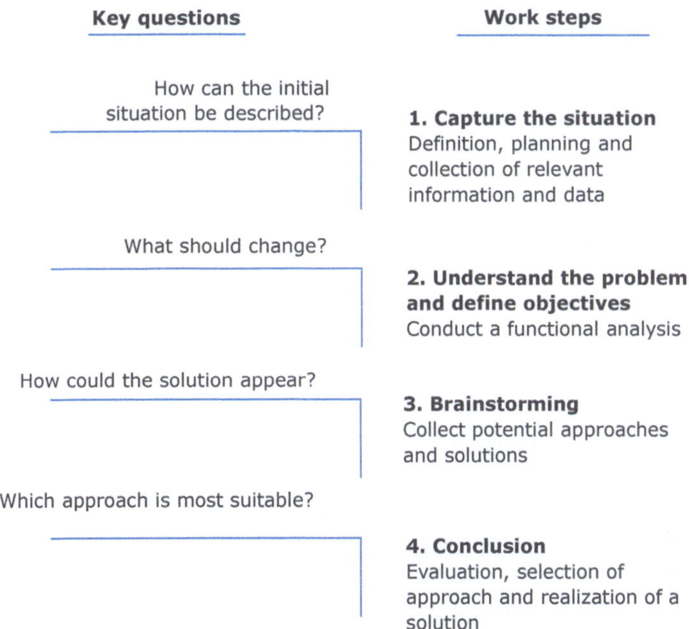

Fig. 9.3 Working steps for value analysis (in accordance with DIN EN 12973 (2020) and VDI 2800 (2010))

9.3.2 An Example of How to Apply the Method

Imagine a door-locking company that wants to expand its product portfolio by offering innovative "smart home" solutions, including a "smart door lock", i.e. an intelligent door-locking system. The goal is to provide hotels, clinics, and businesses with secure auto-mated access which dispenses with mechanical keys by allowing centrally authorized doors to be opened using an app and a mobile device.

The company decides to conduct a value analysis to design this future system. First, the current situation is captured and the problem and its challenges are analyzed.

1. Capturing the Current Situation by Means of a Use Case Diagram

An initial analysis of the intelligent door locking system can be performed using a use case diagram. This diagram visualizes various tasks and responsible persons. The next step is to define goals and objectives for the new product.

The use case diagram for the intelligent door locking system is presented in Fig. 9.4 and is based on the Unified Modeling Language (UML) notation. The diagram establishes a relationship between functions and actors and illustrates how the starting situation is defined for the specific use case involved.

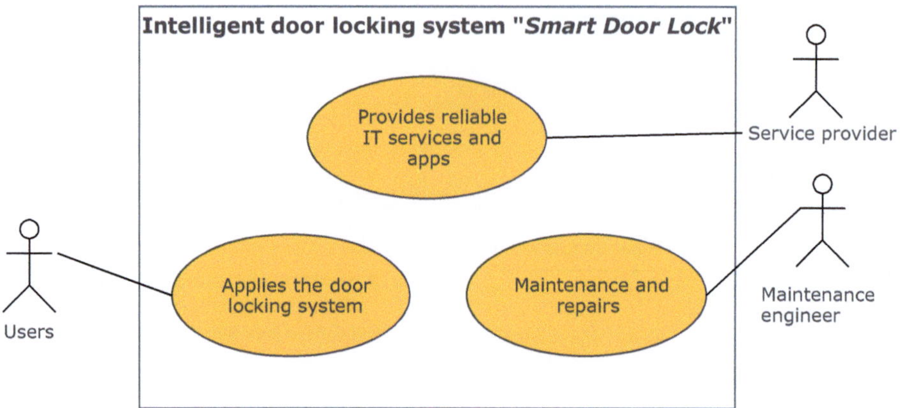

Fig. 9.4 Use case *"Smart Door Lock"*

In this example, engineers develop the technical product that the user will deploy. A service provider offers the IT infrastructure, and technicians maintain and repair the systems.

The requirements for the functions of the automation system can now be systematically derived from the use cases. Detailed functional requirements can be defined for each use case and actor:

– The service provider supplies a reliable application and a cloud platform for managing access rights.
– Maintenance personnel maintain and repair the system and can create new users.
– Users deploy the door locking system for their application.

In practice, these use cases are elaborated and provided in addition to the use case diagram. This documentation is crucial for creating a common and understandable ground for IT development in terms of actors and their tasks. Non-functional requirements which document the quality characteristics of the product should also be described during requirements engineering. One example of a non-functional requirement is the availability or reliability of the cloud platform, something which is central to the locking functions and hence to user acceptance. This approach ensures that the real needs are identified, captured, and understood. The use case documentation describes the starting situation.

2. Understanding the Problem and Developing Objectives by Means of a Functional Analysis

The purpose of this step is to comprehend and analyze the problem. This helps to anticipate product functionality and develop targeted solutions before prototyping starts. The

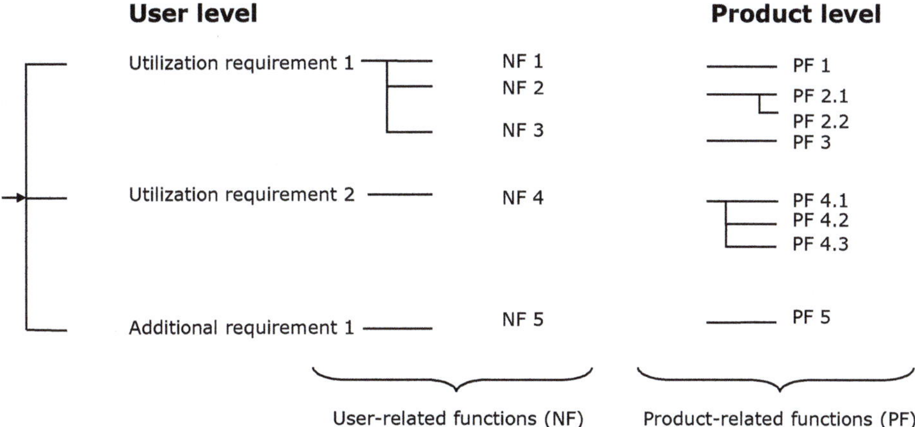

Fig. 9.5   Breakdown of a technical system into user-related and product-related functions

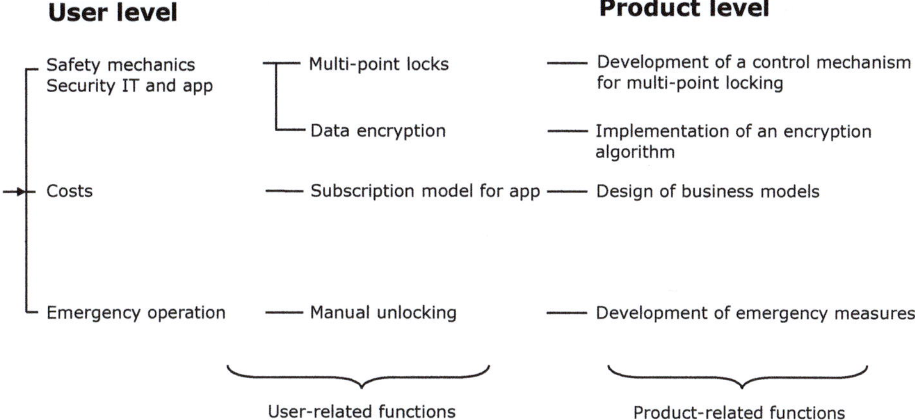

Fig. 9.6   Functional analysis of the intelligent door locking system

functional analysis method is applied in product design in order to tailor products or services to the needs of customers and users.

Functional analysis involves analyzing and developing features and objectives by means of use cases. It is essential to focus on the value rather than on possible solutions. Subsequently, the so-called user- and product-related functions can be formulated.

The concept of functional analysis is illustrated in Fig. 9.5. First, user needs are identified and user-relevant functions defined. User-related functions (UF) are then derived from these requirements and product-related functions (PF) determined. The PFs represents the technical perspective of the product and are formulated for the purposes of the developers.

The functional analysis for an intelligent door-locking system is shown in Fig. 9.6. The following user-level requirements are identified by examining the use case: "door security," "personal data security", "emergency operation behavior in case of power failure" and "cost of the system."

User-related functions are then formulated for these aspects. For example, the incorporation of a multi-point locking mechanism addresses the need for security. The product-related functions are derived directly from these user-related functions, in this case by developing a mechanism for controlling multiple bolts.

Upon completion of this step, the future features of the system are defined from the perspective of the user and the viewpoint of product realization. The interplay between the consideration of user-related and product-related features in the context of understanding the use case ensures that only value-added features are created.

Additional steps:

In Step 3, a brainstorming session is conducted to explore potential technical solutions for implementing the product-related features. A number of different approaches describing possible implementations may emerge from this step.

In Step 4, the "Conclusion" phase, potential realizations are evaluated and selected, resulting in a proposed solution that meets both user needs and technical requirements.

9.3.3   Is there a More Agile Approach?

The methodology described above is standardized in this form and the individual steps are systematically outlined. This has the advantage that a detailed plan is available at the beginning of the work, something which is particularly beneficial for mechatronic products whose implementation requires considerable effort.

In the field of agile development methods, however, there is a trend towards a more flexible or agile approach, cf. for example Cagan (2005). The focus is no longer on the dedicated creation of a concept or plan. Instead, it is on the interaction of developers for the purpose of generating software. While the core methodological ideas are retained, the goal is to improve customer satisfaction through frequent software delivery. This allows feedback from the users to be used in the early stages of development.

Accordingly, the scope of the functional software is considered as the central measure of progress and not the documentation of the development according to the above-mentioned agile methodology. However, agile development requires very close collaboration between business experts and software developers. Value analysis steps are still performed, but incrementally as development progresses.

This form of interaction is much more efficient as problem-solving can be performed in a specific context on the basis of initial software prototypes. However, this approach places high demands on the individuals involved, who must be highly motivated and possess excellent technology development, application expertise, and communication skills in order to convert a lot of good ideas into the development of product functionality.

The development team involved requires the competencies necessary for assessing the application and use of the technical product.

9.4   **From Technical Product to Successful Business Model**

The development of technical systems often requires significant investment and is therefore expensive. For this reason, measures should be taken to verify the future business success of the developments.

As a result, it makes sense to question not only the added value of individual product functions but also the value proposition of the new product. This consideration includes additional aspects contributing to the future acceptance and success of the business model.

The so-called value proposition is built around the products or services to align them with the needs of the customers and the specific applications. Figure 9.7 illustrates three essential perspectives: the user perspective, the application perspective and the product perspective. As Moore (2014) explained in his influential work, these three perspectives are all necessary for ensuring that the product is effectively responsive to the market environment. It is crucial to look at a new product or service in the context of this value proposition. This means considering not only the technical realization of the product but also the user requirements as highlighted in the product design methodology explained above. To ensure market success, it is also necessary to consider the conditions in the application, i.e. in the segment of industry in question.

The example of the intelligent door locking system "Smart Door Lock" illustrates that there are considerations that go beyond product requirements formulated from a function-oriented user perspective.

To be able to realize the product's functionalities during development, it is important to understand the actual customer needs and the conditions prevailing in the market environment, and this understanding should not be overly speculative.

To fully assess a new product, it is essential to understand what value the user derives from the product, i.e., what it offers in terms of "what" ("what does the product offer?"), "how" ("how does the product solve the problem?"), and "why" ("why is the product

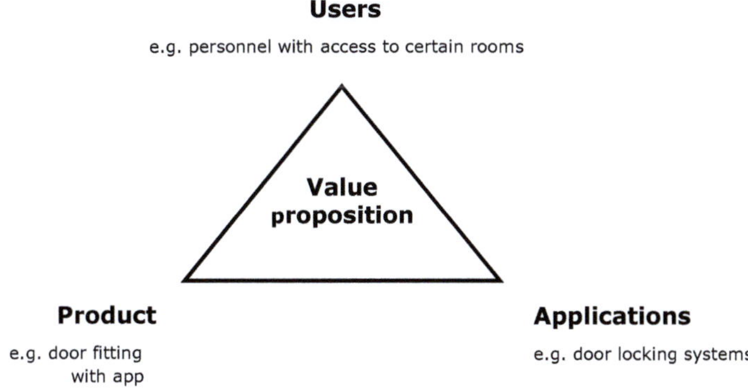

Fig. 9.7 Value proposition according to Moore (2014)

important?"). This understanding is key to comprehending the basis on which value is created.

If we apply these guiding questions to the example of the "Smart Door Lock", we can answer them as follows:

WHAT does the product offer?

– The system can be controlled and configured via an app, which simplifies management in comparison with conventional keys.
– The system meets high security requirements and allows for easy re-keying in the event that a key goes missing or is stolen.

HOW does the product solve the problem?

– The door lock is made of premium materials and can be opened via smartphone apps.
– The system includes a wireless sensor, for example one based on NFC.

WHY is the door lock system important and for what purpose?

– Specific areas require robust protection for various reasons.

The value proposition for the new product "Smart Door Lock" can be formulated as follows:

This intelligent door locking system solution is practical to use and provides secure access to protected areas with many users. Even if keys get lost, sensitive areas can be reliably controlled. The system protects the door from physical and virtual attacks and is in technical terms manufactured from high-quality materials and based on secure software applications.

For example, according to Lukas (2018), a widely used approach today is the "value proposition canvas", which aims to elaborate aspects in a detailed and structured way.

The basic idea of the canvas approach is that customers fulfill tasks using products from a wide range of functional, social, technical, and other categories.

To identify the tasks of a product, the following questions can be asked:

– What tasks are customers trying to accomplish?
– What problems are they trying to solve?

The next step is to examine the obstacles (pains) that prevent customers from successfully completing a task from the customer perspective.

To identify these potential pains, the following questions can be asked:

– What are the customers' biggest difficulties and challenges?
– Where do existing solutions fail to meet user expectations?

Similarly, the gains resulting from a solution are examined. Customers expect to gain from the successful completion of a task.

Gains can be identified by asking the following questions:

– Where does the customer want to save money, time, effort, or labor?
– What can easily solve the customer's problem?
– What is the customer looking for?

A suitable pain reliever describes how pain can be treated. It can be identified by asking the following questions:

– How can the customer's difficulties and challenges be eliminated?
– How can we provide a better solution than other vendors?
– How can negative consequences for customers be avoided?

Similarly, gain creators are examined. These represent what a product does for the customer. Gain creators can be formulated by asking the following questions:

– How can tasks be made easier for the customer?
– How can customer needs be met?
– How can we deliver what the customers want and need?

Table 9.2 contains the value proposition canvas for the locking system. This can be used to create a detailed specification or requirements document. This document provides a clear direction for technical development.

Table 9.2 Canvas for the example of the "Smart Door Lock"

Task	Pains	Gains
– Protection of the facility from intruders – Restricted access – Isolation of particular areas	– Limited functionality in the event of a power cut – Security threats – Data storage and data protection	– Remote access control – Remote monitoring – Simple access control
Products and services	**Pain relievers**	**Gain creators**
– Reinforcement of the door frame – Use of sensor technologies for the reliable detection of authorized persons – Use of heat- and sound-insulating material	– Battery-powered emergency operation – State-of-the-art encryption algorithms – Local storage of personal data	– Keyless entry – Remote locking of the doors in the event of a security breach – Easy key generation

This canvas allows many additional aspects to be considered. However, the new technological approach does not necessarily have to provide an instant solution to all user problems. While it is advisable to embrace new technologies, a critical reflection should also consider potential limitations and counter-concepts in order to maintain a balanced perspective in terms of opportunities and risks.

9.5   Prompts for Reflection

This chapter has introduced some basic approaches to designing automation in a goal- and value-oriented manner. The methodology of value-based system design we used systematically incorporates an understanding of use cases and the desires of system users. Advanced concepts such as value proposition and the canvas approach help to extract the real needs in order to better address the solution of a problem The professional literature discusses additional methods allowing us to better understand the readiness of markets for new products and make assumptions about the potential success of new products.

However, all of these analytical methods and conceptual models are no substitute for engineering skills or entrepreneurial spirit, but instead help us to understand tasks better and approach them in a systematic way.

Set yourself the goal of understanding the value of a new technology and developing a value proposition. Chapter 2 introduced the concept of the digital twin, an idea which has gained significant popularity and has become a key technology in the field of digitalization. The digital twin has the potential to innovate systems in mechanical engineering or create novel networked and intelligent products.

How do you assess the value creation of the digital twin in the field of industrial robots?

a. **Use the following hypothesis and antithesis along with the pro and con arguments to outline the requirements for the digital twin!**

The following thesis and antithesis can be formulated to comprehensively discuss the use case of a digital twin for production system automation:

Thesis: *"The digital twin adds value to industrial robots and machine tools. In the future, the digital twin should be included in all new products, enabling the real-time collection and analysis of machine data based on models to provide new services."*

   Antithesis: "Developing and maintaining a digital twin is an unnecessary investment of time and money without any clear benefits. Since this technology is still in the early stages of implementation, the actual practical benefits are relatively minimal compared to the effort involved."

Table 9.3 provides support for arguments for or against the new technology of the digital twin.

Table 9.3 Supporting arguments: pros and cons of the digital twin

Arguments in defense of the thesis	Arguments in favor of the antithesis
• This is the only way to make reconfiguration easy and fast.	• Creating the digital twin requires a lot of development effort and expense
• New predictive maintenance capabilities enabled by artificial intelligence are provided	• Additional maintenance of the models and their data, as well as the software, will be required
• Adaptations can be carried out automatically (so-called self-x capabilities)	• Incompatibility must be assumed if components from different companies are deployed
• Remote monitoring and control or teleoperation in real time reduces the number of skilled workers needed	• Structure and maintenance require skills and specialist personnel
• The efficiency of the technical system can be improved by optimizing processes	• There are certain limitations due to the confidentiality of personal data

b. **Form an opinion on the following questions:**
   - Is the digital twin an automation technology of the future, or is it a hype?
   - How do you assess the risk of competitors gaining a significant innovation advantage if your company does not take the lead?
   - What are the unresolved questions and issues that need to be addressed in this emerging field?

c. **Conduct research and brainstorming and provide a recommendation for a feasibility study that could cover specific fields of application.**
   - Compare various technology approaches and outline the capabilities of the new system.
   - Develop a use case diagram and list user and product-related functions.
   - Derive your conclusions and evaluate potential implementations that could undergo a feasibility study.

d. **Consider what statements can be made about a value proposition and what information can be included in a canvas against the background of the current state of knowledge.**

Further Readings

Saigal, N.: Elements of a value proposition: A recap from Google's design sprint conference 2018, 2021. https://medium.com/n5-now/elements-of-a-value-proposition-a-recap-from-design-sprint-conference-2018-2b2e393baae2

References

Cagan, M.: **Agile development processes**, Silicon Valley Product Group, Internet Blog, 200523. https://www.svpg.com/agile-development-processes/

DIN EN 12973: **Value Management**; (in German), Beuth-Verlag, 2020. https://doi.org/10.31030/2874697

Lukas, T.: **Business Model Canvas—Geschäftsmodellentwicklung im digitalen Zeitalter**. In: Grote, S.; Goyk, R. (eds.): Tools for Leadership from Silicon Valley (in German). Springer Gabler, Berlin, Heidelberg, 2018. https://doi.org/10.1007/978-3-662-54885-1_9

Moore, G. A.: **Crossing the chasm: Marketing and selling disruptive products to mainstream customers**, 3. Edition, Collins Business Essentials, 2014

VDI 2800: **Value Analysis**. VDI, Beuth-Verlag, 2010

Introduction of New Technologies into Industrial Automation

10

Abstract

Automation is driven by technological innovation and has continually reinvented itself over the past few decades. Automation integrates new technologies and methods to make them usable for industrial applications. The goal of this chapter is to provide conceptual models, backgrounds, and decision points in the implementation of automation engineering for practical applications.

– Why are new technologies introduced so slowly at times?
– How are questions of economic viability evaluated?
– What are the present limits to the use of automation technology in very large systems?
– What can be done to speed up implementation processes and engage stakeholders?

These questions are of course far-reaching, and they can only be touched upon briefly here. However, this chapter offers a taste of the ways in which new technologies can enter the domain and how they might be introduced in practical terms.

10.1   A Look Back

To understand how technologies find their way into industrial automation in practical terms, it is helpful to look to the past. With the benefit of hindsight, trends and patterns useful for assessing the evolution of technologies can be identified. Figure 10.1 illustrates milestones in automation engineering and indicates the changing significance of the disciplines. While automation technology was dominated by mechanical engineering in the

© The Editor(s) (if applicable) and The Author(s), under exclusive license to
Springer-Verlag GmbH, DE, part of Springer Nature 2024
M. Weyrich, *Industrial Automation and Information Technology*,
https://doi.org/10.1007/978-3-662-69243-1_10

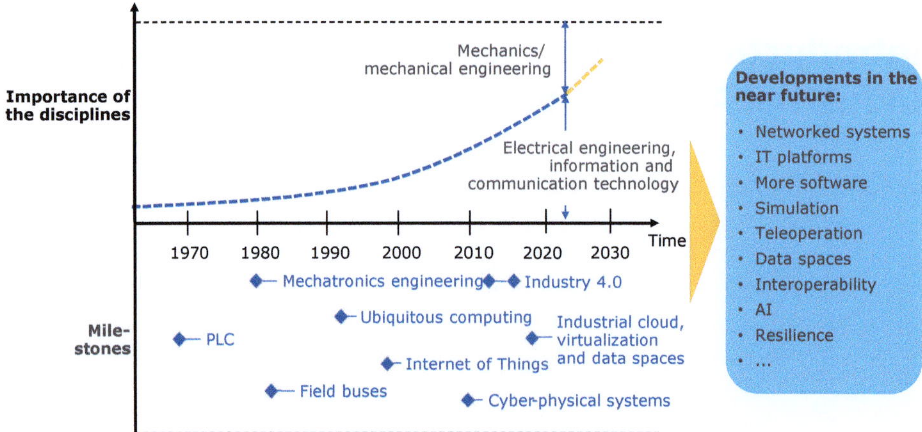

Fig. 10.1 Milestones in automation technology

early days, electrical engineering, information technology, and computer science have now become increasingly important.

With no claims to completeness, the history can be summarized as follows:

In the years from about 1970, individual automated tasks were undertaken on the basis of mechanics and electronics and the first programmable logic controllers.

It was also around this time that the first robotic systems and modern machine tools were developed.

Programmable controllers increasingly allowed fixed wire logic to be replaced by simple process programming.

Microprocessors became more affordable and the first communication networks (such as fieldbuses) were available by the mid-1980s. Mechatronics, which attempts to combine mechanics, electronics, and software in a joint system design, also emerged at this time.

Starting in 1990, software gained increasing significance until it gradually eclipsed mechanical engineering activities. As a result, from about 2000 onwards, information technology and software systems became integral components of technical systems. Since then, there has been much discussion of the pervasive nature of embedded systems and networked software in the context of so-called "ubiquitous computing" and later in terms of the Internet of Things.

At the same time, industrial robots and machine tools have played a leading role for decades and were revitalized by the concepts of "Industrie 4.0" emerging around 2012. In recent years, the strengthening of the "world of information" has played an important role along with connectivity in the sense of linking to data spaces with all of the digital information provided by "real" physical systems.

This look back shows that individual trends last for about a decade while they may not be implemented in the field until two decades after the original idea emerged. These relatively long periods are caused by the time required for the development and operation of automation in production systems and products and by generational changes in developers, who often need time to transition from one development paradigm to a new one.

Looking ahead, it is becoming apparent which new technologies will eventually become important. Increased networked communications, IT platforms, and large software systems are trends that will continue in the future. Teleoperation, data spaces, interoperability, and AI will undoubtedly become significant, even though the practical implementation cycles may be longer.

A new aspect is the desire for more resilience, i.e. the resistance of automation systems to changes and external influences, for example in the field of IT security for industrial systems.

10.2 Selection of Automation Technology and Decisions on Deployment

In the field of automation, many technologies have been developed over the past few decades. Fieldbus systems, for example, have been extensively deployed in practical applications. Why can these technologies not be replaced by more modern communication technologies? After all, in consumer electronics, we experience innovation cycles of only a few years. Why does automation technology continue to evolve fieldbuses instead of developing entirely novel technologies?

The answer is relatively simple: smartphones, laptops, and similar devices have a lifespan of about 3 to 6 years. After that time, these systems are replaced and upgraded to something more advanced. By contrast, in industrial automation, power plants or manufacturing facilities may—depending on the industry involved—have a lifespan of more than 20 years. Similarly, automobiles and other industrial assets such as machinery have in many cases been operated autonomously and without modification for more than 10 years.

10.2.1 Distinction Between the World of Information Technology (IT) and Operational Technology (OT)

The lifetime of technologies and the demands made on their reliability and durability are key criteria that distinguish automation technology from the much more transient world of consumer electronics, for example.

A distinction can be made between the world of information technologies (IT) and that of operational technologies (OT). This categorization is particularly widespread in the field of product automation, but it plays a role in other domains too.

Figure 10.2 illustrates the two levels of consideration—IT and OT—along with their domains of application and the tasks associated with them.

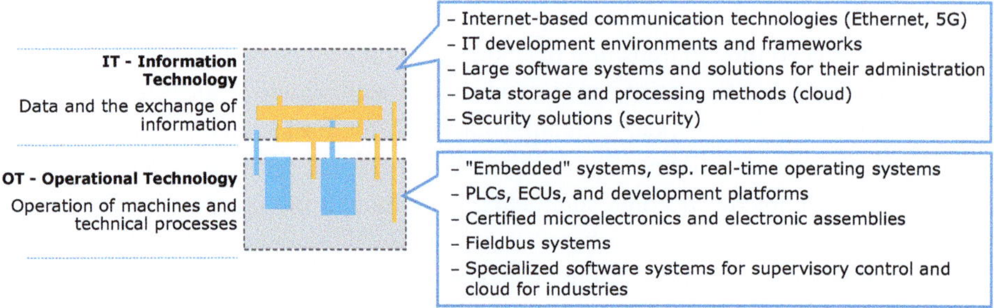

Fig. 10.2 The system environments of IT and OT

The IT level encompasses a company's administrative software, which is organized with the help of servers, networks, databases, and specialized processes. IT refers to technologies that have evolved under the influence of both Internet and enterprise IT and focus on the data-centric implementation of software systems like those used for inventory management or accounting.

OT, on the other hand, stands for operational technologies involving the control and monitoring of events, processes, and devices in real-time. OT refers to systems in close proximity to or of direct relevance to the technical process and designed to be as robust as possible. For example, process control of reactors in a chemical plant is a typical example of OT which must be adapted to the practical needs of the industry using cables and explosion-proof cabinets.

Historically, the differences between OT and IT have provided a clear distinction between the two fields of application. With increasing digitalization, however, these fields are becoming more connected. On the one hand, business processes are being networked with technical systems, and at the same time, IT is being transformed into a high-performance low-cost technology. For instance, cloud solutions for operations management have become reliable and possess real-time capabilities allowing IT to be used for overarching control in the context of OT.

The extent to which technology in the machine-centric domain of industrial automation will converge with widely used information technology cannot be conclusively assessed at this time.

On the one hand, there have been significant advances in IT development. These include the emergence of numerous system architectures and Internet of Things frameworks as well as complex technological capabilities to be found on cloud servers.

On the other hand, the result is an increase in complexity that raises doubts about the suitability of highly complex IT and software systems for safety-critical technical operations.

In many areas, it is becoming apparent that IT needs to be adapted for operational use, implying that it cannot be deployed in its original form. Certainly, IT can be securely integrated with higher-level systems such as ERP or MES (see Chap. 3). However, the question arises as to whether the long-established fieldbus systems used in OT should be replaced by modified IT networks—and this should only happen if the advantages are abundantly clear.

OT and IT will continue to encounter friction at their boundaries until standardization is achieved in the application.

The concept of an "installed base" of established OT technologies already in use supports this view. After all, retraining staff and changing IT system landscapes would require significant effort. However, the use of cloud-based controls from the IT realm may offer cost advantages and make the switch worthwhile.

10.2.2 Reasons for Utilizing OT and IT in Practice

There are significant differences in how, where and why IT or OT is used and which requirements these systems have to meet.

Table 10.1 presents a comparison.

The table for example illustrates that change intervals can vary widely:

People have long become accustomed to updates in office software. However, changes to long-lived industrial controls and similar equipment are sporadic and infrequent. Moreover, the requirements in terms of availability and reliability exhibit notable differences. For example, maintenance downtime at weekends or at night is acceptable for office software. Ad hoc updates to smartphones and laptops are tolerated too. In contrast, automation systems in power plants, heating controls, vehicles and the like must be in continuous operation, and downtime is not an option.

Table 10.1 Differences between OT and IT

	Information Technology (IT)	Operational Technology (OT)
Original area of application	Automation of business processes such as cost accounting or order dispatching	Real-time measurement, control, and regulation of technical systems and industrial products
Life span	Regular replacement every 3 to 6 years at the latest	Use of legacy systems with operating times of over 20 years
Technical modifications	Regular updates, often online	Occasionally, depending on the manufacturer
Availability	Extended office hours, maintenance breaks at weekends or at night, for example	Continuous 24/7 use, high costs during breaks in operation
Security check-ups	These take place; IT security policies exist	Regular testing of the hardware, "security through obscurity" in the software domain, i.e. the specific way in which it works is unknown to attackers

In these cases, the consequences of a system failure leading to a business interruption can be significant. For this reason, technical systems must be highly reliable and consistently available.

Approaches also diverge in security assessments across different domains. A system should not compromise other systems, nor should it allow manipulation from the outside by unauthorized persons. However, the consequences of potential security vulnerabilities for humans and the environment vary dramatically, ranging from mere inconvenience to catastrophic outcomes.

To detect viruses and similar threats, corporate IT regularly conducts security assessments. In contrast, it is still common in industrial automation to only make occasional checks of software on controllers, routers, etc. In the context of process-oriented control software in particular, the principle of "security by obscurity" is used, which prevents attacks by leaving causal relationships unclear.

Process data analysis is now playing a crucial role in production automation. For example, it should be possible to determine the causes of complex error patterns when machines in production facilities are maintained remotely. As industrial systems become networked, new connections are emerging between factories with their automation equipment and the outside world via the Internet. As a result, special precautions must be taken to ensure the safe and reliable operation of automation systems. This is particularly important for elements of critical infrastructure such as power plants, wastewater treatment facilities, and transportation systems.

10.3   Cost Considerations

To comprehend how new methods and approaches in automation engineering make their way into practice, it is essential to understand the criteria that apply there.

From the perspective of the user, the initial focus is on the function or capability an automation system provides. Depending on the industry segment and the application involved, these functionalities can vary significantly. For example, they may encompass the ability to drive autonomously, the provision of information for managing global value chains or the ability to produce goods automatically using mechanical systems.

The focus shifts to effectiveness once this basic functionality has been provided. The emphasis is then on determining the contribution of the automated system to value creation or evaluating potential losses due to downtimes.

The identification of factors that hinder overall economic efficiency and the devising of strategies to avoid losses are often at the heart of the automation technology application. The Overall Equipment Effectiveness (OEE) metric introduced by Nakajima et al. (1988) is a widely used measure of overall plant efficiency, particularly in manufacturing. Figure 10.3 illustrates the three key metrics—also known as key performance indicators (KPIs)—used to determine the overall equipment effectiveness.

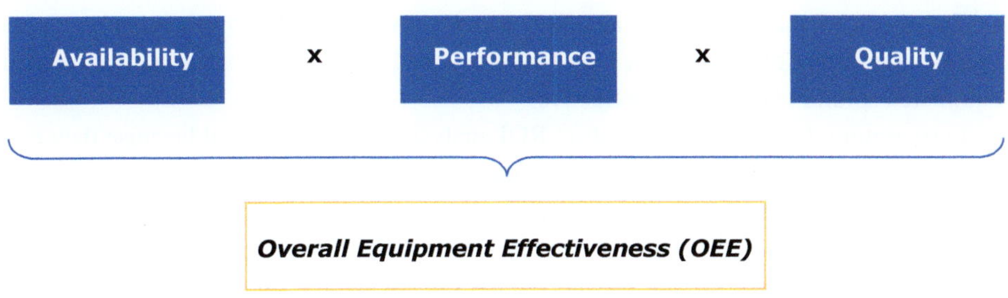

Fig. 10.3 *Overall equipment effectiveness* (OEE)

One hundred percent achievement of the OEE metric is desirable from the manufacturers' point of view.

This involves the following considerations for each metric:

– Availability: Achieving the planned uptime of the automation system is critical to achieving availability. Unplanned technical disruptions, maintenance or repairs should be avoided as much as possible. The same applies for unproductive downtime due to setup or modifications.
– Performance: System performance should not be impaired due to insufficient work or poorly organized operations. It must be ensured that the system is constantly supplied with tasks and is not idle.
– Quality: The quality of delivery is significant too. Faulty manufacturing and deviations from intended operation should be avoided as they cause waste.

The decision to implement automation technology is typically based on OEE metrics. If an analysis indicates that automation can increase the OEE metric, then the use of automation technology is considered worthwhile. In other cases, automation is utilized in order to operate the manufacturing facility more economically than before.

As a result of the increased overall system effectiveness, the return on investment (ROI)—i.e. the ratio of the improvement in relation to the investment—is also decisive in this consideration:

The **return on investment** (ROI) is the key performance indicator for the profitability of the increase in efficiency in relation to the capital invested for the implementation of automation technology.

The way in which the threshold for the return on investment (ROI) with regard to a decision to implement automation technology is measured depends very much on the industry in question. In many areas of manufacturing technology, the maximum ROI for an investment is 2 years, meaning that the improvement in efficiency must amortize the investment within 2 years.

These considerations of OEE and ROI metrics reveal that deploying automation and industrial information technology may not be prudent, especially when a manual or semi-automated solution is more cost-effective.

Furthermore, OEE metrics and strict ROI analysis are controversial because they concentrate solely on the effectiveness of a defined technical system. Aspects such as flexibility or adaptability are neglected or reduced to the quantification of downtime, for example during reconfigurations. This approach places little weight on the agility of the system as the effort associated with future product changes is not difficult to quantify.

Furthermore, without reliable benchmarking data, the added value of implementing IT is hard to quantify.

For instance, if a new system is designed to enable teleoperation, the potential increase in OEE resulting from remote expert intervention may have to be weighed against the related investment, which is challenging to quantify.

This explains why many industries hesitate to adopt new technologies; positive changes are difficult to quantify without an extended trial period. Unexpected failures or challenges caused by immature technology can also reduce OEE, causing unforeseen costs and further reducing ROI.

However, a decision based solely on economic considerations can impede innovation as accurate metrics require a baseline that can only be accurately determined after an investment in automation technology has been made.

To overcome this deployment dilemma, production automation pilots often collect metrics in order to draw conclusions for the entire facility. However, decisions must sometimes be made without the validation of a pilot.

This is especially important when it comes to process control, operations management, and resource management, which can only manifest their effects on a broader scale.

In conclusion, while economic considerations play a central role in the utilization of automation technology, they cannot always be determined with any degree of certainty.

10.4 Technical Limits to the Utilization of Automation Technology

Technical feasibility is an explicit requirement for an automated system, which should be of manageable complexity, i.e. realizable, as well as modifiable. An automated system must exhibit behavior that is predictable and reproducible.

10.4.1 Limits of Feasibility Due to Complexity

As complexity increases, the feasibility of automation systems becomes more important. Practical considerations include questions such as: Can a system be built or modified with a reasonable amount of effort?

Does the complexity of the technical implementation make the automation technology difficult to control or only manageable at considerable expense?

From a conceptual point of view, many approaches to implementation are possible today. In practice, however, pioneering efforts often run up against limitations that result in failing or unreliable deployments. There are no clear criteria according to which the designs of automation systems can be evaluated in terms of their feasibility. However, failed projects from the past have shown that not all of the possibilities which appear to be feasible ultimately lead to a system that functions reliably in practice.

Due to their complexity, the feasibility of the implementation of automation systems depends on the number of inputs and outputs and the number of subsystems involved. This method allows us to at least make a basic classification. Figure 10.4 shows a system of coordinates in which the number of subsystems in an automation system is juxtaposed with the number of inputs and outputs.

The number of subsystems and inputs/outputs can be estimated for typical industrial automation systems. This allows us to get an idea of the complexity of the system.

The assessment for various systems is as follows:

- A simple automated product such as a power drill is an independent system with few inputs and outputs. Manufacturing machines, on the other hand, have multiple controlled axes and various subunits.
- Automobiles have become much more complex in recent years, integrating many individual subsystems (such as engine control, lighting systems, navigation, and driver assistance) with more inputs and outputs.

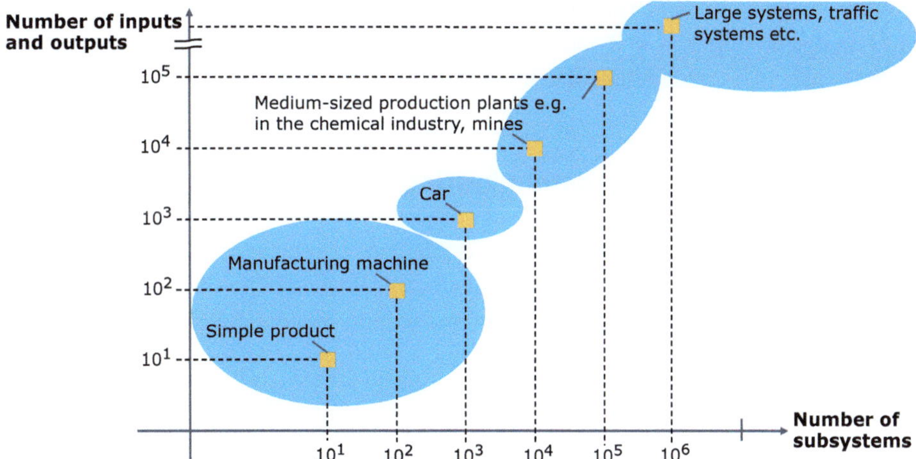

Fig. 10.4 Complexity of technical systems

– Chemical production plants or mining facilities possess many inputs and outputs served by a considerable number of subsystems.
– Traffic coordination systems such as railway traffic control systems operate with more than 10^6 subsystems.

Are there limits to the feasibility of automation systems, and if so, where are those limits? There are in fact practical limitations to the number of systems with centrally controlled inputs and outputs. According to Delsing (2017), these limits for centrally controlled systems are currently around 10^5 inputs and outputs due to the necessity of having all of the information available for system design and implementation at the time of development.

There are also other factors that contribute to complexity, such as network meshing, which can lead to a loss of human comprehension at some point. In addition, systems in the field may evolve incrementally, i.e. they are expanded step by step, requiring the IT architecture to handle a complexity for which it was not originally designed.

The current trend is clearly towards larger systems. System networks in large cities (known as megacities) can have many subsystems, not all of which are conceived simultaneously in the design phase but some of which are added gradually over time. These system networks are characterized by changing system components and their ability to cooperate at runtime. These subsystems are interoperable, meaning that the various components can exchange information and work collaboratively. Control is typically event-driven.

However, challenges arise when the future use of subsystems is not fully determined in the design stage. When components start to interact with each other, new and unanticipated situations can arise. But how can components be designed to be easily "connected" to other components later? The design of interconnected automation networks of this kind involves determining how individual subsystems can be distributed and connected interoperatively to achieve overall functionality through composition or orchestration.

10.4.2 Limits Due to Non-deterministic Behavior

Automated processes must react in a predictable and comprehensible way to ensure system safety during operation and when faults occur. Predictability of system behavior—determinism, that is—is a critical feature of automation systems. Determinism means that a system operates in a specific manner which is understandable and reproducible at all times.

Lauber and Göhner (1999) define an automation system as follows:

> An automation system is considered deterministically if, for every possible state and for any set of input information, there is a defined set of output information and a clear subsequent state. The system responds to identical inputs by yielding identical outputs.

This requirement is essential for conducting tests, inspections, and certifications. For example, when an aircraft autopilot system is tested, it is important to ensure that it behaves in a reproducible manner. This reproducibility is necessary if we wish to understand what happens in scenarios. If a test case yielded unpredictable and varying results each time, the outcome would not be reproducible and the system could not be certified.

From the perspective of developing and testing systems, the requirement of determinism is essential in order to understand and trace the technical function of a system under all circumstances.

However, this requirement is eroding under the pressure of technological advances for two reasons:

- Due to the complexity of technical systems, determinism may be present, but the complexity can create the impression of unpredictable behavior. The effect of deterministic chaos describes a seemingly irregular behavior that still follows deterministic rules. The observer may lose track of how the automation system behaves when multiple components interact, making it very difficult to determine whether determinism exists.
- As automation systems become more and more autonomous, they interact in complex environments and with humans, giving rise to socio-technical systems. Certain questions arise about the human control of autonomous or intelligent systems of this kind. For example, image recognition systems are trained on sample images. There is no simple "yes" or "no" answer to the question as to whether a particular pattern in the image has been detected. Instead, an extensive field test is required, culminating in a statistical statement based on the frequency with which the quality target was achieved. This indicates that absolute statements are not possible and there is only a probability instead, for example one indicating that the test is positive 99.8% of the time.

The demand for determinism is understandable from an engineering perspective, and clear system boundaries combined with predictability are important from a legal viewpoint. However, it is now becoming obvious that the integration of intelligent automation systems that perform self-organizing functions or even learn and react autonomously does not allow a delineation of this kind.

Instead, people will have to get used to the idea that statements about the system behavior of complex automation systems come with probability indications rather than clear deterministic assertions about an obvious system behavior.

For example, while people have long accepted accidents caused by human driver error, the acceptance of accidents caused by autopilot failure in autonomous vehicles is still far from being a reality in terms of product liability.

10.5 Non-technical Challenges Involved in the Use of New Technologies

Quite often, new technologies fail to make their way into industrial automation, even when they have reached a maturity level that would allow for industrial implementation. Developments in recent years show that technological breakthroughs have been missed out on because development trends have been overlooked and companies have missed the opportunity to enter the market. However, new information and communication technologies are leading to new paradigms of value creation. These are changing existing industries to such an extent that delays can result in obsolete products, market share being lost, or participants exiting as competitors.

10.5.1 What Hampers Technological Innovation?

Despite numerous examples of technological leaps in the past, the question as to when and how to introduce innovations is often challenging to answer. This is due to the nature of many technological developments, which often progress in the form of incremental innovations, i.e., in many small steps. However, when gradual changes of this kind cross certain tipping points, a development that has been linear for many years can suddenly turn into a disruptive innovation. This happens when the technology reaches a level of maturity that enables robust deployment over a wide area. Companies are often surprised by these tipping points, either because they did not anticipate them or because they did not prepare for them in a timely manner on the basis of discussions conducted within the company.

This moment marks the triumph of new market entrants who consistently invest in new technologies and are perceived as innovative pioneers. Their products are considered more valuable than those of companies that only incrementally develop existing technologies. As a result, innovative firms can gain advantages when established competitors—the incumbents—lose ground.

However, innovations can fail too, and the causes and patterns of these failures have been under discussion in the management literature for years. First of all, technological developments must be correctly identified and assessed. This requires research and development that evaluates new opportunities in a pragmatic way, provides evidence of feasibility, and identifies practical paths to the future.

It also requires identifying the right time to take decisive action. If the introduction of a new technology to the market takes place too early, there may still be skepticism in the market, and this can lead to a lack of demand. On the other hand, if it is introduced too late, other market participants may have already become established by then.

A problem that particularly affects large successful companies is known as the "innovator's dilemma". According to Christensen (1997), this dilemma arises when incumbents are reluctant to abandon established technological paths in order to avoid relinquishing their existing market. However, they risk being overtaken by new entrants who embrace new technologies.

The reasons are easy to understand: firstly, incumbents resist because the new technologies are immature. Existing markets have little tolerance for experimentation, and incumbents are unwilling to take risks and are skeptical of innovation. At the same time, innovative new entrants will aggressively pursue the implementation of new technologies, gain a competitive edge, and drive incumbents out of the business.

10.5.2 Bringing Innovation to Life

Many companies are aware of this challenge and have measures in place to take advantage of innovation at the right time. However, there is no one-size-fits-all solution as it often involves questions of corporate culture, organizational structure, and stakeholder attitudes, which vary widely and whose actual influence is difficult to assess.

The actions and challenges can be outlined as follows:

- Agile research and development processes are key to identifying, adopting, and implementing innovations. However, these processes depend highly on the individuals involved, who must identify and assess technological trends early on. Innovation, which always requires creativity, experimentation, and a willingness to take risks, is inhibited once a climate of "conformity" has been established.
- Spinning off innovative businesses to independent companies is a common practice used to overcome the innovation dilemma as it removes the dependency on the incumbent company. The new business unit can work autonomously with an innovative corporate culture to develop fresh ideas. However, there are risks too. For example, the implementation may be unfocused or get bogged down in technological details if inappropriate personnel from the incumbent company are involved.
- Establishing innovation scouting with support for start-ups. For large companies in particular, it is important to set up specialized departments that monitor and support innovative companies in order to become aware of new ideas. Feasibility studies and competent personnel can then be integrated into the company's own development. However, cultural differences between established departments and start-ups need to be managed.

Despite these measures, entrepreneurial foresight is crucial. Customer needs and market changes need to be identified early on to allow us to respond to them in time. On too many occasions, multiple innovations and necessary changes have been overlooked, leaving companies to fall behind.

10.5.3 How Do we Engage Stakeholders in Technology Rollouts?

The introduction of new technologies into an organization and the associated questions about opportunities, risks, and costs are evaluated by various stakeholders. These evaluations are controversial at times. It is critical to understand the different stakeholder perspectives (e.g., product design aspects, broader management issues, or commercial interests) if we want to understand who should be consulted about which aspects.

Consequently, it is essential to actively involve stakeholders in the planning and implementing of a new technology rollout.

Leadership behavior also plays a role, as evidenced by extensive psychological analysis. Different types of leaders act very differently and create a climate that is either favorable or harmful to innovation. For example, a "driver" enforces new plans, an "opportunist" merely follows guidelines, a "preserver" tries to maintain the status quo, and a "creative" may set overly ambitious goals.

A look at these characters suggests that technology adoption needs to be managed in a way that encourages and empowers people, allows them to make wise decisions, and leads in a goal- and results-oriented way. Many technology projects have failed to materialize or have encountered difficulties during implementation because the "climate" was not properly managed, creating obstacles that resulted in project cancellation.

The method of stakeholder analysis was proposed by Mendelow (1981) and helps in the systematic collection and analysis of information to clarify whose interests should be considered when developing and implementing projects.

Stakeholders can be classified on the basis of their interests and can significantly impact the development, introduction, and success of a product. Stakeholders can be engaged in activities by anticipating stakeholder expectations and delivering results.

Figure 10.5 explains a classification of stakeholders into four groups organized in terms of importance and impact along the "influence" axis and in terms of participation and engagement along the "interest" axis.

Group 1 includes stakeholders with a low level of interest and a moderate level of influence on development. Their involvement should entail a minimum amount of effort as their evaluations will be of secondary importance. An example of this group of stakeholders would be people from neighboring departments who have heard about the activities but will not be participating in them.

Group 2 comprises stakeholders with a high level of interest but a low level of influence on technical development. While they have no technical influence on development or implementation, their interest can be a source of additional ideas. For this reason, this group should always be kept informed of progress in an effort to benefit from their insights. Examples of this group are financiers and investors.

Group 3 has a low level of interest but a high influence on development. This group of stakeholders can significantly impact the success or failure of the development if they are not adequately involved, making the project vulnerable to third-party influences. For example, a works council or data protection officer concerned about personal data can

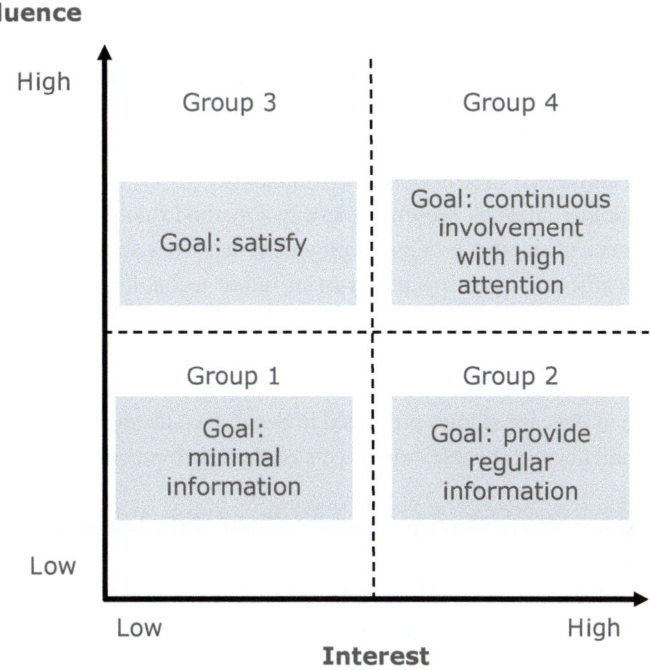

Fig. 10.5 Stakeholders and their classification (adapted from Mendelow, 1981)

create obstacles during implementation and obstruct the project. This group must be consistently satisfied in order to avoid negative consequences for the project.

Group 4 is the most critical group, and it displays a high level of interest and influence. This group should be extensively involved in the development process. It includes top management sponsors, pilot customers, and representatives of research and development partners who shape the project with their assessments.

10.6   Prompts for Reflection

The foregoing discussion has shown that there is a complex array of reasons for or against the use of automation technology. However, economic considerations often play a key role in deciding whether to implement automation. Factors such as increased flexibility or improved quality also play a role but are often hard to quantify in financial terms.

When deciding whether or not to deploy automation technology, implementation issues must be addressed: Should tried-and-tested operational technology (OT) solutions be chosen? Or should new avenues of information technology (IT) be explored? Or is there already a convergence of OT and IT that makes a distinction of this kind redundant?

Discussions about the use of automation technology reveal that investments are often necessary in order to truly realize the added value. However, sometimes strategies are pursued whose benefits have not been clearly proven. In many instances, the pace of development is so rapid that new forms of automation emerge suddenly. A lot of the reasoning for or against deployment often remains relatively vague until the systems have been implemented and tested in real-world environments.

The stakeholder analysis presented here is a method that clearly reveals the attitudes and the influence of interest groups. Deployment decisions should be justified technically and economically, and openness and trust in future technologies are critical for making wise decisions. Ultimately, it is the more or less knowledgeable people—no matter whether they are innovators, conformers, or preservers—who are responsible for whether a new technology is implemented or not.

Select one of the case studies presented in Sects. 10.6 through 10.8 such as the automotive IT case and discuss the relevant aspects of the development of the technology.

a. **How do you evaluate your approaches in light of the issues discussed above?**

What risks does the use of OT or IT have for development?
How do you evaluate the complexity of the technology?
Consider questions of economic viability and the potential impact of technology implementation on OEE or ROI.

b. **Which stakeholders will be affected by the new technology, and what are their attitudes towards it?**

Consider developers, management, vendors, production managers, service experts, etc. Create an overview matrix along the lines of Fig. 10.5 and justify the categorization.

References

Christensen, C. M.: **The Innovator's Dilemma: When new technologies cause great firms to fail**. Harvard Business School Press, Boston, 1997. https://www.hbs.edu/faculty/Pages/item.aspx?num=46

Delsing, J. (Ed.): **IoT automation: Arrowhead framework**. CRC Press Taylor & Francis Group, 2017. https://doi.org/10.1201/9781315367897

Lauber, R.; Göhner, P.: Automation **of processes 1**. 3ed Edition (in German). Springer, 1999. https://doi.org/10.1007/978-3-642-58446-6

Mendelow, A. L.: **Environmental Scanning - The impact of the stakeholder concept**. International Conference on Interaction Science (ICIS), 1981. https://aisel.aisnet.org/icis1981/20

Nakajima, S.; Bodek, N.; Nakamura, K.: **Introduction to TPM: Total productive maintenance.** Japan Institute of Plant Maintenance, 1988

Heading Towards the Future

11

Abstract

Although the path forward is uncertain, the progress and evolution of automation technology can be analyzed with the help of future research methods.

Maturity models that delineate potential development paths are presented in this chapter. In addition, it includes an examination of current trends allowing us to gain insights into the possible future directions automation technology may go in.

The following questions are addressed:

- What path is automation technology currently taking?
- How can the developments be evaluated and compared with the help of maturity models?
- What developments can be expected, and what comes next?

This concluding chapter provides a systematic outlook on future developments, allowing at least a preliminary identification of opportunities and challenges.

11.1 Assessing Developments with the Help of Maturity Models

Enhancing industrial processes and products is achievable by leveraging data and information processing methods to introduce novel functions or capabilities. These transformative prospects are commonly denoted by terms such as "digitalization", "networking" or "AI".

In the field of automation technology, research has produced numerous maturity models that illustrate potential stages of digitalization. While these models do not provide specific guidance on the exact trajectory of further development, they outline a conceivable path into the future and provide a framework for guidance.

These models are based on perspectives that help us to assess and evaluate new technologies.

Three different models are presented below, each illuminating the state of maturity from a different perspective. The first model focuses on the digital maturity of products and is based on an assessment of technology adoption allowing the creation of intelligent and connected products. The second maturity model outlines meaningful evolutionary steps in manufacturing and suggests a sequential approach to digitalization. The third model anticipates stages of evolution in human-machine interaction and illustrates possible degrees of autonomy.

These models provide insights into potential future developments. However, the underlying perspectives have their roots in the present, and it is plausible that these notions may become obsolete in the future.

11.1.1 The Digital Maturity of Product Capabilities

What capabilities can new "smart" products develop through digitalization?

Technologies can be identified and structured on the basis of an expert survey using the Delphi method (cf. 8.3.1) (Weyrich et al. 2017).

This leads to a description of capabilities or skills that characterize novel products which can capture, analyze, and further process data and information with the help of AI.

Figure 11.1 presents an overview of the capabilities derived. They can be divided into two main groups: "Capturing and processing data and information" and "Analyzing, planning, and drawing conclusions using AI".

a. **Capabilities for capturing and processing data and information:**

These capabilities manage the relationship between the system and its environment and enable the system to process data, establish networked communications, and integrate with other information technology and cyber-physical systems. The defining aspects of these capabilities are:

- Data capture capability: This enables the acquisition of large amounts of data. To accommodate the complexity or rapid changes in data, access to multiple databases—either centralized or distributed—should be performed quickly, preferably in real time.
- Data processing capability: Under the influence of the available sensors, the integration of multiple sensor types, and potential sensor fusion, this capability is critical for the system's perception of itself and its environment.
- Connectivity capability: This describes how a system can communicate with others by establishing a communication network that exchanges data and information between different systems.

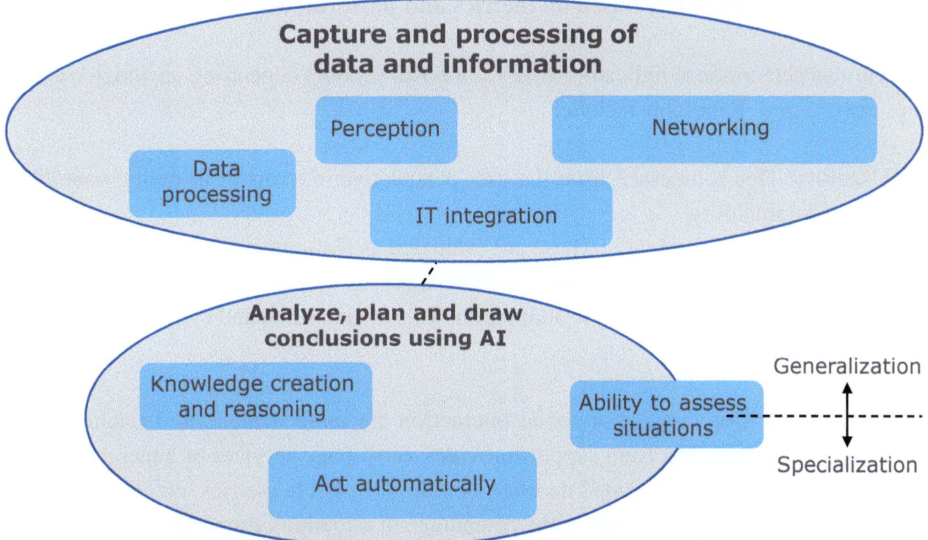

Fig. 11.1 Capabilities of connected smart products (adapted from Weyrich et al., 2017, courtesy of © Springer International Publishing, 2017. All Rights Reserved)

– Interoperability: Each subsystem should automatically integrate into an IT system, in which interoperability may vary in syntax and semantics on the basis of factors such as a unified alphabet or a common language for communication.

b. **b. Capabilities for analyzing, planning, and drawing conclusions using AI:**

These capabilities relate to the perception and evaluation of information and the ability to draw conclusions for the planning of actions. Key elements include:

– Perception capability: This describes a system's ability to generate knowledge from data to recognize its environment.
– Evaluation capability: This involves accessing knowledge in order to understand and trace the causes of a problem.
– Automatic action planning capability: This generates activities such as creating an action plan to navigate an obstacle course.

Also relevant are autonomous specialization and generalization capabilities, which allow a system to specialize in tasks or generalize relationships. It is important to note that, since the number of product capabilities may not be immediately perceived as an advantage by the user, the set of technical capabilities implemented does not directly equate to product quality. As a result, the model serves more as a checklist used to compare and evaluate the features implemented.

c. **Determining maturity based on metrics and performance indicators:**

Various performance indicators arise for a novel product depending on which capabilities are realized. Examples include:

- Usability: This is assessed from the user perspective in terms of usability, operability, and maintainability.
- Modularity and complexity: These are evaluated as features of the system architecture, considering the number of modules and interfaces, as well as measures for nesting them.
- Reconfiguration and automatic adaptation: This is quantified and evaluated for specific situations.

Metrics for decision support or social interaction are more demanding to define due to measurability issues. Decision support provides knowledge services or generates suggestions for facilitating an informed decision or action. Possible metrics include the number of decision situations supported or the percentage of situations supported successfully.

Social interaction, which is considered as an advanced capability, demonstrates the interest of systems in each other, enabling information exchange and even negotiation. Metrics used to measure social interaction could include the availability of language, the use of semantic definitions, symbol recognition, and the ability to learn.

11.1.2 The Digital Maturity of Processes in Industrial Production

The model presented by Schuh et al. (2020) for digitalization in production system automation focuses on the gradual development of system capabilities corresponding to different evolutionary stages. This model defines five steps towards a new generation of self-optimizing and autonomous automation systems.

Table 11.1 illustrates the levels of maturity in the development of automated manufacturing systems for the process industries that are expected to incorporate IT functionality in the future.

Table 11.1 Maturity levels for digitalization in industrial automation (adapted from Schuh et al. 2020)

	Capabilities of the system	Characteristic
1. Connectivity	Collecting and aggregating data and information	System components are networked in terms of information technology
2. Visibility		A digital twin organizes information
3. Transparency	Analyzing and evaluating	Relationships are identified
	Interpreting and perceiving	Information is interpreted
4. Predictive capability	Predictive simulation	Forecasts are made on the basis of simulated scenarios
5. Adaptivity	Artificial intelligence	Self-optimization and autonomy

According to this model, automated manufacturing systems will be networked in the first step, that is, they will comprise components capable of exchanging data. In the second step, data transparency will emerge due to sensor-based data collection. The third step will be to achieve manufacturing process visibility through interpretation, detection, and perception functions. All three steps together can be understood as data collection and analysis.

Further steps for analyzing this data which follow the principles of artificial intelligence are conceivable too. The interpretation and perception of the data can occur at different levels of complexity. A predictive simulation that projects multiple scenarios into the future can be used for forecasting (step 4). The fifth and final stage (step 5) is a fully adaptive production system (Adaptivity). Future systems of this kind will be self-optimizing or even autonomous.

The model does not describe how and in what form these maturity levels are achieved. However, it outlines a typical chronological sequence (collection, networking, and data analysis) geared towards the realization of individual capability domains rather than the evolution of production system automation. Maturity levels may be skipped, for example, when insufficient data is available to input a "what-if" simulation for the purpose of predicting the future.

11.1.3 The Evolution of Human-Machine Interaction Towards Autonomy

With the increasing deployment of AI and the implementation of autonomy skills, control can be gradually passed from humans to autonomous systems.

Figure 11.2 illustrates the maturity levels on the journey from assistance to automation and further on to autonomy, based on Parasuraman's et al. (2020) framework.

On level 1, the human makes decisions independently; on level 2, a selection of options is presented. On level 3, the system narrows down the options, and on level 4, it suggests an action to take. One example of this could be an assist system for traffic lights that reacts to an expected traffic jam by suggesting an alternative sequence of traffic lights to the controller.

At this stage, while the human may choose to do otherwise, the system waits for confirmation on level 5. Then, subject to a veto period on level 6, the system executes the action fully autonomously on level 7. Systems of this kind are in use today for well-defined tasks and known system boundaries.

From level 8 on, control shifts to an autonomous system that manages its relationships independently. On levels 8 and 9, the human may still be informed of execution, while on level 10, actions are organized without any human involvement whatsoever.

This explanatory model is primarily concerned with the automation of decision making and the selection of options for action with regard to the possibilities of human intervention. The classification is based on the interaction and the results of the system.

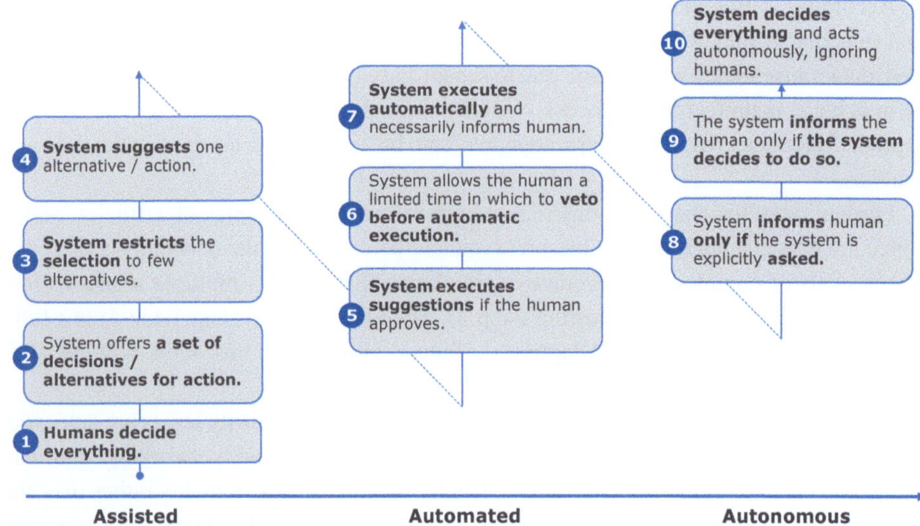

Fig. 11.2 Interaction levels of assistance, automation and autonomous systems (adapted from Parasuraman et al., 2020)

The boundary within which an automated or autonomous system operates is essential. If the entire system is clearly defined and delineated, such as in the regulation of room temperature by a thermostat, it is more likely to be considered as an automated system because all conditions of use and control were foreseeable in the design stage. Autonomous systems, on the other hand, navigate a partially known world. In this "open world", an autonomous system can navigate and even explicitly modify its environment.

11.2   What Technological Trends Prevail and What Developments Are Foreseeable?

The preceding discussion has made it clear that there are numerous directions in industrial information technology from which automation technology can benefit in the future. Within this context, several trends that will shape the development and business of automation in the years to come are now emerging.

Information and communication technology as well as computer science can be expected to lead to numerous innovations in the networking of systems and in data handling, human-machine interaction, and artificial intelligence.

The extent to which one development will prevail over another and lead to significant changes in practical terms cannot be predicted with any degree of certainty. Widespread adoption depends on a range of variables, including practical acceptance, added value, support from platform vendors, success in standardization, and others.

A retrospective view reveals a continuous evolution that can also be disruptive due to breakthroughs in specific fields of technology. While disruptive innovations of this kind are understandable in retrospect, the diverse nature of automation technology makes them difficult to predict for the future.

11.2.1 Trend: Ubiquitous Connection of Systems in the Field

The interconnection of automated systems in the field with the engineering departments of manufacturers offers completely new options for modifying software during operation, for example by means of updates, and for gaining insights into actual operating conditions based on the return of sensor information.

One evolutionary step in system development has been underway for some time now: Over-the-air updates provide software upgrades for automated systems while they are running. Today, the implementation of software engineering ranges from structured processes and software process improvements to continuous integration and deployment (CI/CD). The development processes of cyber-physical automation systems already use a CI/CD loop with automated software delivery, requiring new IT architectures such as microservices and containers. Connectivity through 5G/Edge and new standards for testing different software versions such as ISO/SAE 21434 (2021) play a critical role too.

Data feeds with operational data from the field are considered a key to success in developing AI systems in automation: In a data loop, AI is trained and validated using continuously collected field data. The data loop has already been introduced in the case study on automotive architectures (Chap. 8). Data generated by systems in the field is fed into the AI. Given the development requirements, a selection of scenarios from the real environment, often enriched with synthetic data, must be made in order to provide an appropriate training base for the AI.

The trained AI modules must then be validated using additional scenarios. After successful validation and testing, an update allows the improved AI capabilities to be deployed in real systems.

The future data loop can be used to develop autonomous systems for scenario-based learning for AI, using real field data and additional synthetically generated data for AI training.

Development via the data loop should focus on specific operational scenarios in the Operational Design Domain (ODD). The nature and scope of the ODD is critical for the learning behavior of the AI. ODDs of this kind can ensure the compliant operation of the autonomous system based on the trained AI because they represent the officially specified use cases for which the AI should function. The ODD essentially specifies the operational scenarios for which an autonomous system must be designed. This for example includes the environmental conditions in which an autonomous vehicle can operate. The ODD defines a type of use scenario and, if necessary, scenarios for non-intended use.

New types of tests based on the ODD—see for example Thorn et al. (2018)—and used in the field of autonomous vehicles are therefore of particular importance as they define use cases in the sense of test-based development.

11.2.2 Trends of Modularity, Connectivity, the Digital Twin, and Autonomy As Well as AI

The current developments in the field of automation aim at a comprehensive use of software and in particular the exploration of the information and data space. Moreover, AI will lead to major changes in autonomous systems. The changes are mainly driven by the modularity of automation components, connectivity in terms of information technology networking, data and information modeling in the digital twin, and the use of AI for control and guidance in autonomous systems.

These technologies form the basis for connecting software and systems engineering to provide new agility through continuous development (DevOps), networked software platforms, and the systematic use of data and information. AI for the perception and planning of autonomous capabilities is already playing a critical role in pilot applications and empirical evaluations.

The four technological fields mentioned above contribute various characteristics, some of them essential:

a. **Modularity and interoperability of automation components.**

Low-cost ECUs are driving the modularity of components, allowing automation systems to be built from modules in place of using single solutions. This modularization allows the development of "intelligent" components that can be easily combined. Modules simplify the development of automation systems because the entire system can be orchestrated from components. Interoperability of modules is ensured by standards, thus allowing easy interchangeability. Standardization and the availability of modules from different manufacturers enable the long-term operation of systems. This saves development time, makes software complexity manageable through encapsulation, and ideally reduces costs by using off-the-shelf products. The increasing complexity of automation can only be managed using modularity.

b. **Connectivity for information technology networking.**

Connectivity, i.e. the connection between all components and modules, is essential for the smooth flow of data and information. Technologies such as real-time wireless technologies (like 5G) and enhancements to Ethernet using Time Sensitive Networking (TSN) are evolving rapidly. Communication with overlying, underlying, and adjacent systems, their components, and people is improving in terms of latency, data rates, reliability, and the IT security of data and information transmission. To avoid data islands, the integration of communication partners should take place without extensive engineering efforts to ensure the adaptation and "translation" of data and information. Systems with standardized interfaces and languages must be used to exchange existing data.

c. **Digital twin for the virtual representation of technical systems.**

Data should be digitalized, stored and continuously updated from the beginning, not in many different systems, but always in one or more digital twins instead. A digital twin can be realized differently to enable a digital representation of the real world in an information space which maps properties, states, and behaviors of objects and processes. Algorithms for implementing AI-based services can then be built on top of this. The result is a data space in which data and information are fully or partially represented, from the development and production of the individual system as well as assembly and commissioning to lifelong operation and recycling. Data and information can be evaluated at any time and used for new software services and improvements. The digital twin leads to virtualization, in which the location of software execution and data storage is technically irrelevant; they can be hosted remotely in a cloud or evaluated in a networked embedded system.

d. **Autonomous systems based on AI.**

Autonomous systems and AI are key enabling technologies for the future. AI can automate tasks which were previously impossible or only possible with a disproportionate amount of effort. If AI is consistently used for control, automation systems will not only be able to follow predefined program structures but will also be able to make decisions independently in the future. In the years to come, technical systems are expected to have components with autonomous capabilities that can be combined to form complex overall systems. This will lead to new capabilities for automated systems to assist humans or take over control, even replacing human decision-makers. The economic and societal benefits will reach dimensions that match or exceed previous stages in the development of automation and digitization. Breakthroughs in AI leading towards self-learning systems only hint at the future. Examples include DeepMind AlphaStar, which reached the grandmaster level in the strategy video game "StarCraft II" in 2019, or ChatGPT, the AI chatbot based on a large language model that has been making waves since late 2022.

11.3 What's Next?

This book outlines the driving fields of technology, the technologies of the future, and the cross-industry portfolio of topics to be found in automation technology. As an integrative science, automation technology ensures knowledge transfer to practical applications.

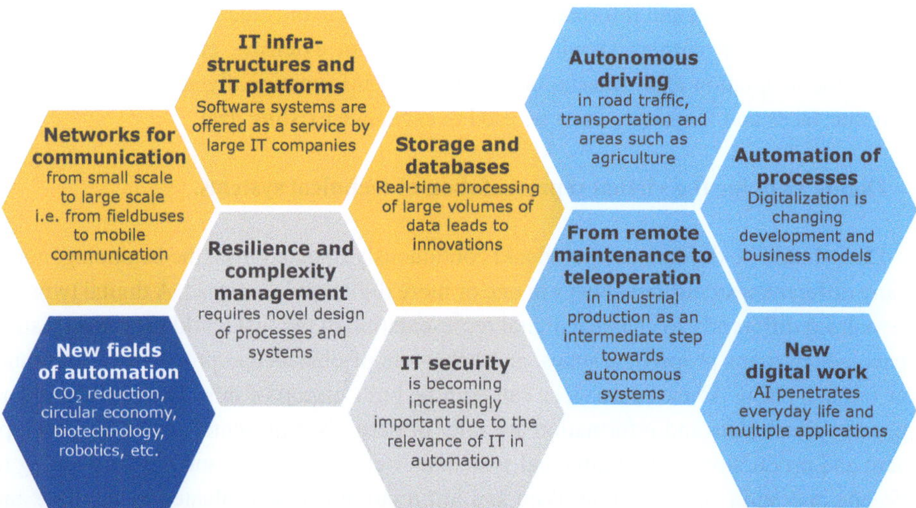

Fig. 11.3 Key topics of future automation

Current topics are extensively researched, tested, and evaluated in practice in a discourse involving academia, manufacturers, automation technology suppliers and IT vendors as well as consulting firms.

Many of the developments follow an incremental course, evolving over a number of years and finally arriving at a practical implementation. However, there are also several technologies that exhibit disruptive behavior due to technical breakthroughs.

An analysis of observations from conferences, interviews with innovators, roadmaps, etc. reveals the topics shown in Fig. 11.3.

Technologies related to communications, data, and information networking as well as storage are particularly important from the perspective of the industries involved. These technologies require significant investments in IT infrastructure to be deployed on an industrial scale.

AI for autonomous systems is primarily seen in a number of application areas. Significant investments are currently being made in autonomous driving, the automation of industrial development and business processes, and workforce digitalization. Activities related to IT security, building resilience, and the management of increasing complexity are also receiving significant attention.

Entirely new applications of automation technology are also expected to play a critical role, such as innovative ways to reduce CO_2, produce hydrogen, or support circular economy initiatives. It can be a challenge to find the right balance between the opportunities and the risks. Enthusiasm for innovations that lack maturity can lead to disillusionment among users, while inactive companies risk falling behind technologically and losing their competitive edge.

The pace of the adoption of new technologies and innovations varies significantly across industries. Some industries have strong reservations about the value of technological innovation, fearing high implementation costs and substantial risks. On the other hand, industries such as automotive or home automation can excite users with technological possibilities, and this often makes economic cost considerations secondary as the focus is on introducing products with new features.

It is already foreseeable that future waves of automation will significantly change the working and living environment, directly affecting many more people than today. This raises many new questions about the trustworthiness and reliability of autonomous systems which are closely linked to ethical and societal considerations.

By the end of this book, readers should have gained an understanding of the opportunities and risks of automation and industrial information technology, thus providing them with insights, reflections, and practical examples. The responsibility for taking up these challenges, driving developments skillfully, making informed decisions, and taking the right action to shape a sustainable future for automation technology is now in the hands of engineers, technologists, and decision-makers—in your hands, in other words.

Further Reading

Deichmann, J.; Doll, G.; Klein, B.; Mühlreiter, B.; Stein, J. P.: **Cracking the complexity code in embedded systems development.** Company Magazine McKinsey, 2022
Gartner: **Emerging technology roadmap for large enterprises 2021-2023.** Company Magazine, 2021
Tilley, J.: **Automation, robotics, and the factory of the future.** McKinsey, Firmenschrift, 2021
Wahlster, W.; Winterhalter, Ch. (Hrsg.): **German Standardisation Roadmap Artificial Intelligence.** English Version 2, DIN und DKE German Commission for Electrical, Electronic and Information Technology. 2022

References

ISO/SAE 21434: **Road vehicles. Cybersecurity engineering.** Beuth, 2021
Parasuraman, R.; Sheridan, T. B.; Wickens, C. D.: **A model for types and levels of human interaction with automation.** IEEE Transactions on Systems, Man, and Cybernetics - Part A: Systems and Humans, Vol. 30, No. 3. 2020. https://doi.org/10.1109/3468.844354
Schuh, G.; Anderl, R.; Dumitrescu, R.; Krüger, A.; ten Hompel, M. (Hrsg.)**: Industrie 4.0 Maturity Index—Managing the Digital Transformation of Companies.** Acatech Study, München 2020.
Thorn, E.; Kimmel, S.; Chaka, M.: **A framework for automated driving system testable cases and scenarios.** Report No. DOT HS 812 623. Washington, DC: National Highway Traffic Safety Administration, 2018
Weyrich, M.; Klein, M.; Schmidt, J. P.; Jazdi, N.; Bettenhausen, K. D.; Buschmann, F.; Rubner, C.; Pirker, M.; Wurm, K.: **Evaluation model for assessment of cyber-physical production systems.** In: Jeschke, S.; Brecher, C.; Song, H.; Rawat, D. B. (Hrsg.) Industrial Internet of Things: Cybermanufacturing Systems, Springer International Publishing, 2017. https://doi.org/10.1007/978-3-319-42559-7_7

Index